Digital Production Resources Management

By Theophilus Edet

Theophilus Edet

 theoedet@yahoo.com

 facebook.com/theoedet

 twitter.com/TheophilusEdet

 Instagram.com/edettheophilus

Cover design by **Benedict Edet**

Copyright © 2022 Theophilus Edet All rights reserved.

No part of this publication may be reproduced, distributed, or transmitted in any form or by any means, including photocopying, recording, or other electronic or mechanical methods, without the prior written permission of the publisher, except in the case of brief quotations embodied in reviews and certain other non-commercial uses permitted by copyright law.

ISBN: 9798413398654

Table of Contents

Preface 5
Why Digital Production? 9
BOM Formulation 26
Routing Composition 42
Work Scheduling 61
Production Execution 79
Production Planning 98
Production Management 116
Production Order Queuing 136
Production Routing Management 147
Work Center Management 157
Material Management 172
Maintenance Management 189
Accounting For Production 213
Infrastructure For Production 254
Groomed Production Systems 268
Production Related Digital Skills 287
Digital Production Strategies 312
Fourth Industrial Revolution (4IR) 331
Global Economic Growth Prospects 344

Preface

Production has gone totally digital, device-driven, and smart. Digital Production Resources Management is a resource handle that portrays manufacturing through stratified processes and resources of machines, manpower, margins, materials, and methods as the foundation for modern, flexible, smart, best in class, digitally driven production operations.

Digital Production Resources Management conceptualizes the sequencing of digital production through four processes of BOM formulation, routing composition, work scheduling, and production execution as described here in this book with dedicated initial chapters. Thereafter, the reader is taken through the more important aspects of digital production such as the people relationships required, the infrastructural requirements, and in fact the various mechanisms associated with the stated four processes.

This book is about automation of production sequences for the transformation of materials by masterful engagement of the resources of machineries, manpower, margins, materials, and methods into finished products, of course digital transformation as the main driver.

The book starts out by describing the process involved, then the mechanisms, people, and infrastructural support. Subsequently we delve into certain manufacturing systems successes and global statutory guidelines to aid manufacturing through computerisation and policies.

This book shows that everything Production in a digital world can be confined to the four processes of BOM formulation, production routing composition, work scheduling, and production execution. Besides, digital production is not only limited to MRP and MRP II but to critical and flourishing handles of demand chains, supply chains, and warehousing where ERP holds sway and into all relationships and resources of the enterprise, particularly customer and vendors relationships.

With Digital Production Resources Management in the bag as an absorbed skillset what's in it for all?

Digital production resources management done right empowers sustainable and flourishing manpower engagements so that manufacturing is not conducted in isolation. As dwindling production yields continue to threaten sustainable livelihoods around the world, action is needed from governments, private sector leaders and international institutions to ensure the continued viability of SMEs to sustain livelihoods everywhere towards responsible consumption and production which is critical to be mainstreamed as priority targets in global development strategies.

Unfortunately, much of the world's economy remain isolated and on lockdown from the rest of the knowledge economy as they readily fall flat on major indices with no appreciable digital readiness figures. These isolated economies can no longer continue to take a hit from avoidable ineffectiveness of past decisions with rising global uncertainties, the time to join the booming world economy is now to register the ambience of an embracing global economy for the best of economic buoyance to enjoy robust value creation through the virtual world of e-business.

Theophilus Edet

Why Digital Production?

Modern society depends on the transformation of raw materials into finished goods. Management of the transformation of raw materials to finished goods in a digital way, cloud based or paperless is the bedrock of modern production. Production is related with goods and manages resources to achieve the transformative objectives for delivering finished goods. Particularly, at this time in global history that manufacturing requirements are beyond the availability of natural resources, demanding manufacturing production to push beyond the envelope of natural resources into the realms of new materials to meet demands for consumer goods.

Sustained economic growth is to be realised with prevailing circumstances of outsourcing production to low cost countries in the global drive for distributed production and sustainable consumption. Such new entrant countries delving into the realms of modern production do not need to follow the infrastructural outlay of the industrialised world to enjoy these gains. However, by following the monolithic path of digital innovation new entrants into the global production climate can quickly gain traction and deliver stunning results in production with key digital tools that plug seamlessly into global supply chains to create new demand chains in the process, achievable through containerization of manufacturing, with possibility to rival the promises of Industry 4.0 treasure chest of the industrial world as heavy capital outlays in the acquisition and programming of robots is not a prerequisite to delve into the world of digital production for the underdeveloped world, most of who are the custodians of natural resources the world needs to venture into manufacturing production.

Production resources are a clustering of machines, manpower, materials, methods, and money, otherwise considered the 5m, for the transformation of materials into finished products. The formulation that produces the finished products need to be formulated in a bill of materials (BOM) that structures the material content make up that goes into making the finished product. To mix all these together manufacturing requires routing schedules providing step-by-step procedures, durations, costs, and machine stop definitions to be operated and followed in the transformation of materials into finished products. Because the variables that go into production is enormous the information content of production resources are better computerised for effectiveness and the monolithic structure of digital workflows.

Management of the transformation of raw materials to finished products in a digital way, cloud based or paperless opens the enterprise to economies of scale. With ingredients of e-business integration of Master Production Schedule (MPS) Material Requirements Planning (MRP) and Manufacturing Resources Planning (MRPII) fully amalgamated, digital production is able to drive all enterprise functions for maximum efficiency and effectiveness, ensuring that high quality goods and services are delivered on time, in the right quantity, and to the right people without exception. Ensuring that all enterprise resources and relationships are sustained and achievable with digital production resources management.

Master Production Schedule (MPS)

The MPS is a framework for integrated material and capacity planning. As a material-based planning tool, the MPS is able to galvanise requirements for production as a digital tool that time phases production and requirements to enable planning enjoy the insights it needs to make things work. As a fundamental tool for manufacturing intelligence MPS plans for finished products based on production capacity requirements. MPS can also make plans for make to order and make to stock manufacturing production with good manufacturing insights for manufacturing efficiency. What the MPS does is to make plans for independent materials which are finished products that need to be produced based on customer orders or manufacturing needs for satisfying MRP requirements. Customer orders from the demand chain are consistent marketplace demand issues that push orders consistently to the conversion cycle where the process of value addition is engaged towards fulfillment of orders as well as manufacturing needs.

The MPS is a critical input to the MRP that explodes requirements for dependent materials that are required to sustain manufacturing resources.

Material Requirements Planning (MRP)

The MRP system began with computerised databases that maintained inventory, away from manually tracked bin cards that used to be prone to error. By shifting to a computer system, companies could make this information available to the purchasing department, where it could be readily consulted when determining how many additional parts to purchase as requirements for producing a given end item. In addition, the data could now be easily sorted and sifted to see which items were being used the most (and least), which yielded valuable information about what inventory should be kept in stock and what discarded. When inputted with an MPS and a BOM the MRP system explodes the BOM into material counts required to deliver production orders, the

MRP also reports a material netting for order release, and offsetting factoring as well as timing such as labour time and machine time required. The work order explosion gives an idea of materials to be back-flushed from inventory when each finished product stated on the MPS acting as input to the MRP is delivered.

Manufacturing Resources Planning (MRPII)

The MRP II system progressed manufacturing computerisation beyond just MRP that is material oriented by incorporating a production schedule (MPS) and a bill of materials (BOM) for every listed item to achieve manufacturing excellence by making plans for manufacturing resources required to consume and transform input materials to finished products. Being an incorporation of MRP, MRPII makes plans for manufacturing production based on resource availability and not just on material availability. This approach balances out manufacturing across capacity plans by smoothing out production to avoid bottlenecks, such as manpower assignment and machine sequencing considering with machine capacity taken into consideration, in the production activity.

A lot happens when production resources take on a digital makeover. Since production resources such as machines, manpower, materials, methods, and finance have all been computerised in considerable depth. Payroll systems computerise the work rates and work timings around manufacturing manpower while inventory control systems do the stock control around materials, accounting systems manage the costing around manufacturing. It therefore follows that the production function itself must of necessity be computerised for digital effectiveness to access economies of scale.

Enterprises that give a digital makeover to their production function stand to deliver a make-to-stock manufacturing environment with no stock-out or backorder scenarios due to the lock step of the master production schedule (MPS) with their MRP incorporated in a watertight MRPII system for best in class material management and manufacturing resource excellence.

Following closely with material efficiency is seamless integration of digital marketing, demand management, procurements, and logistics with production.

Today's infrastructural support, institutional frameworks, and statutory guidelines emphasise attention to digital production to align with ideals of multilateral trade agreements, globalisation, climate change, and standardisation. As such Digital production is a priority for all enterprises engaged in any form of manufacturing to deliver sustainable consumption and production patterns through efficient conversion cycles while strengthening value chains in the process.

The workforces of former industrial milestones are waning out. The emerging workforce is tech savvy and looks forward to an iPhone moment in manufacturing production. Therefore today's production environment must rely on digital interfaces for prominence. Not paying attention to digital production may be detrimental to the delivery of a smart workforce being the responsible of every future proof enterprise.

Innovation and disruption is a factor in today's world of economic growth. Prominent among them are digital disruptions and every manufacturing environment need to digital proof their enterprise to service the needs of customers who flow with digital suaveness into enterprise corridors. Enterprises must go Digital to stay relevant. Such is the cultural and economic climate today that enterprises must plug into the digital socket of manufacturing production to gain significance. Digital is the most important keyword in manufacturing today.

Today's industrialisation levels still assume Gutenberg like importance just as it was in 1450 with all the mechanical attachments, that the society must sustainably produce to support its population. Just as it was in the days of Gutenberg, production must assume the toga of "Produce or perish" as it was with the academic toga of "Publish or perish". A society not producing is therefore perpetually dependent on the producers who in turn control the economies around their consumers if by chance they assume industrial monopoly. Every society needs to produce what they consume and consume what they produce. Therefore, there must be an institutional and policy pressure with commensurate incentives to get manufacturers step into digital manufacturing production for global economic sustainability.

All types of manufacturing production in practice today makes for the industrial scale application of digital technology to augment an already mechanised and somehow automated sequences of production to the extent that digital dashboards in computer terminals rather than wooden panels conspicuously positioned in shop floors become the platform from where manufacturing production commands are marshalled to fulfill the requirements of digitally declared production plans. All manufacturing at whatever scale go through the sequences of material preparation, processing, finishing, and packaging.

Continuous Production: The plant resembles one huge machine in which materials are taken in at one end of the plant and the finished products are continuously despatched from the other end of the plant. The plant runs for 24 hours a day and never stops. Plastics, papers, glass, and boards are produced this way. Digital production is useful in continuous production for estimating longer time buckets of production using digital means so as to maintain larger material accumulation in silo storage, and automation of material handling machineries as some materials here may be injurious to human handling, and an ERP to integrate all relationships and resources of the enterprise.

Mass Production: This manufacturing system plant is laid out to produce a single product (and limited variations of that product) with minimum material handling. The product is moved from one operation and/or assembly station to the next in a continuous, predetermined sequence by means of a conveyor belt system. Individual operations are frequently automated. Such plants usually manufacture consumer goods such as cars and household appliances in large quantities in anticipation of orders. Digital production is useful in mass production to automate the flow of processes by use of supervisory control and data acquisition (SCADA), and inventory control systems with virtual requisition to guarantee the process does not shutdown due to absence of critical materials.

Batch Production: This involves the production of batches of the same or similar products in quantities ranging from hundreds of units to several thousand units. These may be to specific order or in anticipation of future orders. If the batches of components are repeated from time to time, this method of manufacture is called repetitive batch production. General purpose rather than special purpose machines are used and these are usually grouped according to process. Digital production is useful in batch production for automation of inventory, activity based costing (ABC), BOM management, and managing the routing roll-up to reach the limits of production.

Job-Shop Production: A job shop is a kind of manufacturing process that makes small batches of make-to-order kind of production to customer requirements. Job shops require specialised manpower to operate that understand specific routines to operates machines. Digital production is useful in job shops to support scheduling of materials to the floor, manpower assignments, intricate planning to deliver workable delivery dates to customers, and virtual requisition to limit cost of materials.

Right now the world is greatly influenced by smartphone interfaces that with minimal learning almost everyone can use smartphones of every flavour and even the multi-language variants as well. The technology that will be global will therefore have to follow in that same vein of smartphone interface, that favour moving production to computer screens, and that includes tablets and smartphones, being web driven so that there is no interruption due to software updates as updates are applied automatically.

Digital transformation of manufacturing function empowers corporate talents to work smarter, faster, and better. The organization-wide cross function provided by a digital production system through its encapsulated ERP processes enable enterprise integration of best practices and performance coordination, projecting the enterprise to its utmost performance while sustaining its commercial resources and relationships with flexibility that help the enterprise adapt and thrive in global economic corridors.

Transition to digital production has become very popular. With the prevailing scale of global consumption it will be hard to get it right with traditional manufacturing strategies. With the fourth industrial revolution gaining traction customers demand more agile manufacturing environments meaning that manufacturers must immediately move into digital production to stay relevant and make digital production a key element of product lifecycle management to support the creativity and speed needed to stay afloat and competitively differentiated. Digital production is essential for marketplace relevance of manufacturing enterprises. The manufacturing enhancement platform that digital production offers not only creates economies of scale but also scales empowerment of talents to sustain economies beyond present day contagion induced economic disruptions.

Digital is the transformation the world needs to address the frequent disruptions in global supply chains so that with distributed digitally driven productions the world can attain sustainability in the consumption of consumer goods so that tensions can be relieved of depleting natural resources to confront the realities of climate change, such that value chains can be sustained.

These are the days of digital disruptions. Enterprises now have no option but to be digitally oriented. The intelligent enterprise therefore gains insight and increased efficiency for effective response to future digital emergence to secure the sustainability of their enterprises where digital technologies are adequately exploited. The robust new business models that emerge from embracing digital is ever more promising.

Efficient Resource Management

To administer the manufacturing production processes of products for rapid customer fulfilment and warehouse replenishment accordingly requires that enterprises resources and relationships be efficiently managed to increased work centre, machine, and workmanship throughput by digitally streamlining the production processes, for increased profitability realised by reengineering and adjusting production processes for quality control.

Reengineered digital manufacturing production procedures yield increasing efficiency and consistency throughout the production process. Getting it right digitally presents an awesome opportunity to develop and engage the right talent for manufacturing production. For the first world, manufacturing talent is not a problem when prevailing demographics are taken into account but for developing countries it is significant to develop manufacturing production talents digitally and these can open a corridor to virtual production resources to plug every idle resource into virtual work center resources. That is where social manufacturing is going. The capacity to put every idle resource from a vast population into productive use and challenge the supply chain to respond fast enough to the demands of a rapidly changing world by unlocking the paths to sustainable development.

BOM Formulation

Manufacturing production involves the transformative consumption of materials under the conditioning leverage of machines operated by specialised manpower successively executing specified sequence of documented methods to deliver finished products. Since materials constitute the primary inputs into manufacturing production they are better understood in most manufacturing settings for interface to manufacturing resources in production as a bill of material (BOM) being an input of constituent materials of product structure being fed into production for transformation into finished products. BOM is a structured breakdown list of material composition input into finished goods that provide material description and unit quantities required to make one unit of finished product. The BOM helps the production framework express a methodical approach to product development and acts as a call up manifest for product build up in the step-by-step transformation of materials into finished products.

BOM is an inventory of materials and quantities needed to make a product. Their formulation entails deep product knowledge and a readiness to constructively engage manufacturing production in the research and development, making, fabrication, or assembly of a finished product.

First Steps To Manufacturing Production

For production to take place materials have to be formulated as a key input to be transformed by production resources. BOM provides that structure that enables the logical provision of material flow as an input for manufacturing resources to work on. However, the formulation expresses only the material requirements for just one unit of end item and depending on the volume number of end items required an arithmetic juxtaposition supplies the total quantities of materials required as BOM explosion giving how many of each unit of component material is required. Manufacturing programs such as master production schedules (MPS) that provides count quantities of independent materials feed its output to material requirements planning (MRP) systems to realise what is called a BOM explosion where total material counts required for the entire production are expressed, usually in the presence of other manufacturing resources such as production orders and on hand inventory positions held.

BOMs are required to set up manufacturing framework for the production of finished products. They are essential in production that no production can be anticipated without a BOM. BOMs are very fundamental to production that there were times when they were closely guarded secrets, or even legally patented or trademarked. Nevertheless, research and development facilities continue to develop new BOM variants to give their patrons the marketplace edge they so earnestly crave for by analysing, synthesising, and formulating new manufacturing BOMs and reengineering the supporting work methods for their economic ownership for mass production of finished goods by releasing new strains and versions of BOMs to progressively drive product development to the realm of the ideal BOM for better marketplace performance through research and development (R&D).

For economic reasons majority of vendors on the supply chain impose restrictive minimum order quantities (MOQ) for their finished products implying that customers must be sure of capacity for materials management before ordering. The MOQs safeguard the vendors on the supply chain from excessive handling costs as ordering costs are manageable from specific MOQ points at which the vendors operate make-to-order manufacturing economically. This is because every BOM must see beyond prototyping into marketplace sustenance if the finished products they represent are of economic order quantity (EOQ) significance. The BOM is a base functionality for setting up production resources for manufacturing in the digital age and its usage traverses R&D, engineering, product development, marketing, and production.

Fundamental Element of Manufacturing Data

Modern manufacturing is knowledge and data driven. Modern manufacturing computerises such strategic functions of the enterprise as finance, inventory, manpower, payroll, warehousing, supply chain, demand chain, marketing, production, purchasing, vendors and customers. With an adequately computerised inventory system manufacturing operations delve into areas of computerised master production schedule (MPS) that is fed as an input into material requirements plans (MRP) which also takes the strategic input of the BOM alongside the MPS output to generate calculated output of BOM explosion and other outputs of netting for phasing out orders for materials, and then the material offsetting output. No detailed manufacturing takes place without a BOM, and until materials line up in warehouse storage systems during planning manufacturing cannot be executed. BOM is the fundamental element of manufacturing data presented as a subset of inventory data instantiated and encapsulated for the purpose of giving expression to products from conception through design, into production and finally into the marketplace. A BOM after formulation must be tracked to determine product integrity all through the product life cycle through batch production tracking and deliveries, and if modifications and upgrade through product evolution is anticipated it is the BOM that acts as the primary driver for product reengineering.

Every manufacturing operation requires a BOM from where material components are pulled into the product mix during material transformation to finished products. This structured declaration helps to control the flow of materials into the assembly line or work centers where they are fed to be worked on with machines and power tools to eventually produce the desired finished products in the approach documented.

A BOM could contain just one material or as much as possible number of materials and could be printed on as many pages as possible, but as fundamental data they convey the expression of what is needed to produce just one item of finished product, their generic or specialised description, in what quantities, and at what unit cost being a sum of cost of individual components that make up the BOM and when cost estimated the value of a BOM is the direct material cost of a unit of finished product referred to.

BOMs are absorbed by MRP systems to regulate and optimise scalable material sourcing, procurement and replenishment for production resources. MRP takes in the MPS and BOM as inputs to calculate a BOM explosion that provides inventory of exact quantities required for production as gross requirements. This then regulates material provision flow to production from inventory where the bulk of materials are held in the warehouse.

The manifest of materials conveyed by BOMs form the basis for production planning. In planning an MRP system takes the BOM in conjunction with the MPS as inputs to output a BOM explosion that gives the quantity of each material required for production on a certain date. With such plans at hand material procurement from the supply chain could be expedited and held in inventory before that exact date where they are then released based on resource requirements on the specified production date.

As key element of manufacturing data cost accountants hold firm a cost effective BOM to enter into production as the most profitable in terms of materials and in conjunction with R&D are able to substitute alternate items depending on market intelligence to release product variations as base level, mid market and full options products for strategic advantage of the enterprise with respect to desirable market penetration plans.

Design offices, R&D, and customer experience departments hold particular BOM with which product concept is maintained, they frequently thinker with BOMs at this level to observe various materials, alternate vendor capacities, and marketplace options with respect to product ideas. It is based on BOM formulation that products within the enterprise are idealised and released for market dominance within the competitive scope of the value chain for the competitive advantage of the enterprise.

By holding the BOM structure in computerised relational database either in a network or in the cloud on the internet and feeding it as an input in conjunction with the master production schedule (MPS), and the inventory database then the computer can generate requirements for parts, sub-assemblies, and raw materials to facilitate production. This computer based approach or digital production is the bedrock of modern manufacturing worldwide and it signifies an iPhone moment for all manufacturers, as it is a near impossibility for traditional manufacturing to meet up with the scale of digital production as manufacturing could be brought to a halt due to disjointed material plans, more so the status of a manufacturing BOM.

Production planning holds static BOMs specifically approved by the board and management for manufacturing. At this stage of realisation the manufacturing BOM has a leg in the inventory for material planning and another one on the supply chain to expedite procurement based on the material requirements exploded from the material requirements plan (MRP) to sustain virtual requisition.

Therefore, the creative representation of this fundamental data that defines product structure for the creative delivery of finished goods compels consumers to march in droves to storefronts for the acquisition of finished products that came out from the most creative BOMs out there.

BOM is the factor that evolved modern manufacturing through computerization, into Material Requirements Planning (MRP) and then into Manufacturing Resource Planning (MRP II), as gradual development of computer systems that were designed to bring the advantages of computerization to manual manufacturing systems in existence before the 1960s. It began with the creation of databases that tracked inventory, evolved into managing the supply chain, and eventually to running every aspect of enterprise function with Enterprise Resource Planning (ERP).

All these utilise extensive databases and it is mainly the BOM, being the product manifest, that they track to give manufacturing a facade of strategic prominence in manufacturing environments leading to the eventual emergence of Enterprise Resource Planning (ERP). With MRP, MRP II, and ERP manufacturing can now be considered very robust and can handle all aspects of an enterprise, including customer relationships, financial resources, production resources, talent relationships, vendor relationships, and warehouse resources. Besides, ERP can even grow with the enterprise. With ERP the world is sure of decent work and economic growth drive going forward to 2030 as a strategic goal of United Nation's Sustainable Development Goals (SDG) goal 8.

Supports Control, Monitoring, and Tracking of Materials

As a subset of inventory data the BOM is a guide to specific materials that must be held in inventory to sustain production. The BOM helps maintain these as controlled items that must be encumbered for production in inventory as their shortage could lead to production shutdowns. Due to how unique BOM contents are, special inventory storage sections need to be segregated for them to enhance monitoring and control for availability. Their movements in inventory transactions need to be tracked to prevent conflicts and exceptions in their usage so as to deliver good market value to the enterprise.

The way BOMs are configured accords material handling guidelines on the production process from the planning stage due to the condition of the materials from warehousing where they are issued. The composition of materials entering production because of their state requires that there be guidelines for their safe handling for value retention and efficacious value addition. In particular there need to be guidelines for safe handling of chemicals, where BOMs incorporate chemical compounds, and with manufacturing subscribing ISO 9001 this is where the deft application of ISO 9001 finds remedy. It is therefore important that BOMs that incorporate harmful substances must subscribe to a more rigorous framework for safe handling so that eventual product delivery is safe to use. Therefore, documented directives to sustain safe handling, and avoid unnecessary exposure to chemicals by any route that may discharge toxic chemicals must accompany harmful substances.

When BOMs are used in manufacturing there will be an associated yield factor in terms of scrap rate as not all material input will yield 100% production output due to material handling inefficiencies, material defects, and manpower skill deficiencies. Scrap rates must be stated on the header declarations of the BOM structure records. Adding scrap rates into material requirement calculation of MRP runs make for more accurate estimates of material requirements from the MPS.

Non End-Items Could Also Use BOMs

It is not only finished products that use BOMs, product families with sub-assemblies and complex parts need to be associated with BOMs to understand their structure for maintenance, requisition, replacements, and reconfiguration. Sub components of machines often break down and rather than incur excessive downtime, the damaged part could be subjected to corrective maintenance to return the machine back to operation in less time and less cost by efficient management of spare part manifest as a BOM variant.

BOM presentation of materials constitute dependent items of inventory, their consumption dangles on the relevance of the independent item finished product, which could even be an equipment used in production, or in enterprise service. When such equipment ceases to be in operation the independent items do not need to be kept in inventory any longer, and as such inventory cost cutting strategies could be engaged.

Marketplace end items also use BOMs to support their proper functioning and service life. Marketplace BOMs are generally considered as spare parts so that they support the end items as dependent items such that a damaged component could be pulled out and a factory supplied mint used to replace the damaged part by technicians. Such plug-and-play scenarios deliver limited success due to lack of appropriate tools, machineries, manpower, and factory conditions. However, manufacturers, in their bid for better customer experience despatch skilled manpower as field service personnel with the right tools to support such complex parts in a manufacturer developed work center where the right conditions are observed to deliver optimum customer experience in the service of their products.

How BOMs Are Formulated

BOMs are need-based structures. Their targeted creations are for problem solving, and as marketplace solutions they must be formulated with deftness of skills and knowledge. In manufacturing for example BOMs form a major aspect of product design and planning. Before embarking upon a BOM formulation the design of the product must be well understood in great detail. Hence BOM creation must be tailored to needs of the environment before deciding the level of detail to subject the BOM creation and the requirement details.

Next, ensure the BOM is based on materials at hand or else it will be considered a phantom BOM and the materials will first need to be understood after they have been successfully ordered and acquired, and the right amount of details need to be included in the BOM structure.

Then start including details of materials that are well understood, the description, and quantity required for one unit of end item, the unit of measure and the base cost. This must be done iteratively till all the materials required in the BOM are captured.

Double check the information supplied and ensure that they conform to established requirements. After which the BOM structure is frozen and safeguarded as a baseline structure. In future it could be revised but the first one must remain intact with possibility of being available for reference.

The future of BOM is defined by the competitive advantage it confers its promoters who involve their intellectual property in a marketplace hazard to deliver market-changing products and must as a result earn competitive advantage in the process if not making it to being a market leader. The leverage a game changing BOM brings to the market demands enterprises maintain secured access to BOM records likely stored in the cloud, and that appropriate revisions lead to an up to date BOM that is intelligent, smart, and structured, leading to a robust product life cycle, greater accuracy, improved visibility, and manageable margins.

Routing Composition

Conception of production to be carried out in a manufacturing facility need to be composed into routing sheet bill of activities as expression of operation sequences to describe the route to be followed for each step of the manufacturing process as well as resource requirements for consumption in the transformation of materials to finished products to define each step of the manufacturing sequence as a manifest to schedule and engage resources for production activities. It provides the correct sequence to be followed in material transformation so that quality guidelines can be adhered to and that resource configurations, manufacturing execution, and inspections conform to expectations.

The articulated expression for transformation of materials must therefore be exhaustive as to the resource mix that enters the transformation matrix ultimately resulting to finished products. Routing composition determines the sum of operation duration for making just one unit of a finished product, that when multiplied by the volume quantity of finished products required by the production order determines the exact duration of a planned production order that conveys production mandate. Routing sheets list the sequence of procedures required for manufacturing a product, indicating the work center for each operation and instructions required to carry out the work.

Actual production process sequence must be modeled into routing statements to create pathways for the movement of materials through various work centers, machines, and operations, describing resources to be consumed in the process and through designated work centers. Routing is a set of information detailing the method for manufacturing a product. It includes the activities to be performed, the activity step by step sequence, the various work centers to be traversed, and the standards for machine setup and run. The routing manifest also includes information on tooling, manpower skill requirements, inspection, testing, and quality requirements. Most importantly, routing sheets could go as far as specifying time required and cost of each operation step.

Therefore, while BOMs determine materials to be consumed in production per product unit, production routing sheets determine the activities and time required to carry out production as well as the manpower that carries out the operations. Selection of machines and methods of production are composed on routing sheets to ensure appropriate machinery and methodologies are employed in the transformation of materials to finished products. Elements that go into the routing composition include such prime resources as machines, manpower, materials, and ancillary resources as activities, duration, and cost. Routing composition summarises the total duration required to make one unit of an item in time units and will usually be multiplied by the quantity volume of items required to be produced when attached to a production order to estimate the production order's cost and duration roll up.

Production method detailing by the specified operations, sequential treatments, work center travel, setup and runtime requirements, together with manpower requirements is a great boost to capacity utilisation of manufacturing facilities, helping to adequately utilise manufacturing resources in the best possible ways.

Routing evaluations help to clarify the correctness of BOMs, the adequacy of manpower, the workability of machinery, and the adequacy of methods as spelt out for production of specified items. Organisation wide, routing evaluations help to form an integrated production optimisation framework for the robust scheduling of production workflow.

Routing sequences are the instructions that give directives on how to use the BOM whose component materials are to be transformed and consumed in the production of finished products. Decision points and loops in routing sequences help production manpower pay attention to measurements, inspections, and evaluation requirements of the work environment to ensure the routing profile provides decision support mechanism even at such lower levels of production so that the production workflow through the routing sheet is a consistent problem solving mechanism as well.

In traditional production environments there are process plans, routing cards, or routing sheets. However, what they all have in common is production execution sequence no matter the deftness of documentation. The routing sheet provides the guiding path and sequence of operations across work centers through its ordering steps that showcases all activities to be performed on the work with particular emphasis on how they are to be performed in specific work centers, the material requirements involved per routing step, the manpower requirements, and the work duration on each machine stop over makes production routing one of the most important articles of modern digital production resources.

Production routings are methods and instructions detailing how production is to be carried out in routing steps. Production routing composes the step by step mapping procedure required for the manufacturing process in production unit work centers that the job will pass through while being transformed. It provides the exact locations of various machines required and mechanisms to be followed in the transformation of materials to finished products or sub assemblies. Routing specifies the work center stopover sequence the job must achieve to become conditioned as finished products.

Routing Machines transform materials into finished products; they are grouped into work centers. Work centers specify measurable parameters such as scrap rate, cost centers, and manpower resources required as well as setup and running costs required to be committed to keep production running. Machines specified in work centers are defined to be operated by specialised manpower for the transformation of materials into finished products. The specific sequence that moves materials between work centers are outlined in a routing composition so that the journey map of materials across work center locations ensure that appropriate machines deliver the required operation on the specified materials. Machines perform industry specific operations on materials, such as blending, grinding, polishing, painting, and even boxing. Their setup for operation consumes time and financial resources and their runtime also consume financial resources as well. Machines must be loaded within their capacity and duty cycle to avoid breakdown consequences on the flow of production that require maintenance operations to get them up and running. Every routing step will specify a work center where work is to be carried out just by selecting the required machine involved in that routing step.

Routing Manpower entries specify the specialised skills, knowledge, and abilities required to handle and transform materials within the work center and may in the process make use of specialised tools, consumables and personal protective equipment (PPE) as accessories for safety. Industrial manpower is often grouped into crew formations to assure their availability, which are further justified through training and skill accomplishments for organisational efficiency. Manpower usage in production encumbers financial resources that must be accounted for as payroll liability through as piece-rate work requirement in the job specification that sets up the manpower requirements in the routing composition and shown in the routing cost roll-up.

Routing Material Consumption is drawn from the BOM attached to the production order. Material requirements are provided for in the routing composition by drawing from a related bill of materials (BOM) specified as encapsulating all materials required for producing one item unit that the routing relates to. The BOM is to be consulted for the gross requirement to produce specified quantities of finished products. In the routing composition material requirements are specified for each routing step as they are drawn out of the BOM together with handling requirements to make for progressively increased item makeup while being transformed. Materials with which production is to be effected are composed into BOMs. These BOMs specify quantities of each material required for each unit of product. BOMs can be created for finished products or sub-assemblies. Sub assemblies are end items of work in progress operations. The BOM specified for the finished product or end item attached to the routing is only a declaration but the actual material movement is controlled by the material requirements emanating from a production order where the routing composition is will be attached as these materials must be moved out of inventory to execute routing steps as specified in the routing sheet during actual production. The routing dictates when components from the BOM are to be issued to the job as the process of value addition progresses through the work environment. Accounting for the materials could be at material issue, but in today's digital transformation approach a more advanced form of accounting is dictated with back-flushing of materials at completion of production or specified production break points, so that when a production order is completed or reaches a break point all materials involved in the production of the specified items are specified as moved out of inventory by decreasing their on hand quantities as well as crediting the asset accounts of those materials to reduce their account balances.

Routing Activities

Activities describe the work that needs to be done on materials at each routing step to make them achieve the state or condition required for specified purposes along the material's transformation path. Most activities require material conditioning or exertion of force. Thus transforming them into specified material properties as required. For activities to be effective power tools are to be provided to ensure that human effort and capacity are not unnecessarily exhausted or depleted in the process of manufacturing production for manpower sustainability. Methods specify the procedural operations to be carried out to effect production as well as details of activity timings. By establishing operation routings, specific methods can be specified for each production process step. Operations with their groupings and timings are compiled into activity templates to aid future composition of new routings.

Routing Duration

Duration specifies how much time to be consumed at each routing step by specified manpower to perform the stipulated activity that possibly engages machines, consume materials, and conditions item properties in the process. Every routing step must have an estimated duration that must be subjected to constant revision for optimization. The duration is expected to capture the estimated time required to complete the activities, procedures, and methods specified for the routing step. Each routing will normally specify setup duration and operation duration. The setup duration specifies the time required to setup the specified machine for operation, while the operation duration specifies the time required to carry out the specified activity. For only one item passing through the machines the throughput is expected to be a sum of the setup and operation durations but for multiple quantities the throughput is expected to be setup time plus the sum of operation time for all units involved. This type of economies of scale are what makes manufacturing flourish for most enterprises as their deftness in seeing the big picture gives them advantages of scale in being able to eventually fine tune operations for the reduction of wastages, materials, and time requirements for faster throughput thereby making big gains from such ineligible aspects as setup time and time optimisation savings.

Routing Conversion Processing

Routing compositions contain exhaustive information about the product to be manufactured with details of each operation to be performed in transforming materials to finished products. The mode of material consumption, item charges to cost centers, cost template attachment, and machine usage definitions. Routing steps provide the activities and detailed sequence of tasks to follow in manufacturing the product.

Routing components help attach materials from a BOM to the routing step to support transportation or material handling of work in progress into production areas or work centers so that prescribed work is applied in completing the product. All routing operations to be performed on materials in the process of transformation to finished products or sub assemblies must be exhaustively documented for future research and as learning points for product development along the path to product maturity.

Planning Production

Production planning is where routings are defined and held in store for anticipated production operations as may be required. Routing specifications should include activities, production resources, and costs enabling the routing to charge accounts and consume machines, manpower, and material resources when executed. Attaching routing sheets to production orders help to plan modern manufacturing production. The routing structure gives an estimate of time with which production can be completed for the specified product. Besides, the time taken by each operation can be subsequently tracked and reported to help drive the limits of productivity.

Modern production is planned with routing sheet attachment to production order, such that every production order is accompanied with an attached routing sheet to convey defined procedures for consumption of materials from an equally attached BOM as well as to specify the consumption of production resources. One routing sheet should be meant for each work center stop of the production order. For effectiveness, each work center having its specifically profiled route sheet different from those of other work centers enable the specialization of work in production facilities such that planners can dedicate composition of routings to specialised subject matter experts in the production planning team who create, update, and authorise the specifics of operation sequences and resource requirements for manufacturing production on routing sheets that accompany production orders to shop floors and work centers.

Resource Definitions of Machines, Materials, Methods, Manpower, & Money Within Routings

A collection of specialised machines and manpower required for the transformation of materials to produce specified products simply identified as work centers is the jacket in which work cultures persist for the transformation of materials that they receive based on the prevailing routing sheet under the control of skilled manpower. Work centers specify measurable parameters such as scrap rate, cost centre, and skill set required as well as setup and running costs required to be committed to effect production. Work centers define machine pool and production capacity. Each work center specifies the kind of operation it can handle and can receive work orders that engage its resources to effect production.

Activities and cost templates enable rapid setup of routing sheets by applying defined activity profiles with a corresponding cost profile that enable rapid realisation of operation sequences and cost roll ups for setting up required routing sheets for production.

Routing Step Definition on the routing sheet defines intermediate step sequences required in production from material handling to boxing the product. Properly defined routing gives more accurate cost and duration of production route to be taken.

Routing Decisions for Work Center Efficiency

Production routings standardise manufacturing operations to round off wastages and place workmanship in shape to appropriately deliver all the time, ensuring proper throughput for manufacturing operations and work center efficiency to deliver production mandates, accommodate materials encumbered for product transformation and report quality, yield, and scrap status of production operations.

Engrafting Outsourced Procedures

Not all routing sequences must be performed in house. There are situations where an external vendor has to complete some operations on materials and return them back to be finished in house. Such works are farmed out with a purchase order generated from the routing sequence and moved to the next sequence of production when received. However, transporting the materials involve additional handling and freight costs, which must be documented and accounted for in work in progress (WIP) accounting via entries specified in the routing sheet when outsourced.

Routing Pacing Duration

Each routing step includes specified duration times. For a routing sheet the sum total of the time specified for each operation step is the routing pacing duration being the time taken to produce one finished product based on the specified routing sheet. This duration is critical for estimation of time required to complete a production order when taken as a product of pacing duration with order quantities.

Routing Cost Roll-up

Every composed routing is a template for the consumption of production resources of machines, manpower, and materials. Each step of a routing is specified to consume these components while progressively adding value until the end item is derived. Routing cost roll-up for every composed routing is the computation of the cost of manufacturing the end item stated on the routing sheet which are derived by the process of adding direct material, direct labour, and overhead costs of each routing step to obtain the total manufacturing cost per unit based on the composed routing.

Routing cost roll up helps to estimate the cost of production and by deftness of innovative strategies can help to pinpoint where production cost could likely be drained and by innovating around those processes through a change of materials, machine, manpower, methods or even training, a cost reduction strategy could be realised that helps the enterprise achieve reasonable cost reduction and savings.

Standardisation, Sharing, and Exchange of Routing Statements

The following steps show how to compose production routings into routing sheets for use in work centers:

1. Develop production objectives

2. Develop a set of task lists capable of meeting the stated objective

3. Determine production resources needed to implement the stated tasks

4. Create a timeline for a unit of production as a sum of all duration for each task.

5. Determine inspection, tracking, and assessment methods to check and determine that the procedures were carried out successfully and that production adhered to quality guidelines.

6. Finalize the strategy

7. Distribute to all work centers that have the capacity to handle the procedures described for crew learning, workshops, and optimisations before going live.

Work Scheduling

Work scheduling strategies deliver the plans and priorities that lays out production workload captured in production orders as workflow mechanism to convey desired manufacturing activities to be executed on materials as they enter manufacturing facilities. Scheduling secures crew shift and related resources for material handling and transformation across work centers under the guidelines of specified routing sequences required for doing the work specified to release finished products into the demand and supply chains and progressively moderate total production required from a production facility. Market expectations need to be reliably qualified to schedule work, inventory control need to be up to date, there must be availability of machinery and manpower to drive production activities, routings must be well documented and available to provide step by step material transformation guidelines, machine and skillset requirements, and risk factors must be well understood and adequately contained to derive a production plan on which to schedule production of finished products.

The benefit of effective work scheduling is that work progressively and dynamically absorbs production resources without causing congestion to work centers in production facilities based on the stated capacities of production resources that are important in absorbing work schedules effectively.

Production Order Creation

Manufacturing depends on a framework that establishes order out of the chaos of stock and order based demand generated from a master production schedule (MPS) by establishing queues of production order sequences to produce independent items from work centers resourced with suitable machines, manpower, materials, and methods, to carry out the work required for material transformation and value addition to deliver finished products.

The distinguishable make to order (MTO) and make to stock (MTS) production orders approach production facilities differently. MTOs arise from customer orders from the demand chain and its production is for known customers with encumbered funds on the revenue cycle so expenses generated by the conversion cycle into the expenditure cycle are matched to funds captured in the enterprise revenue cycle and as such there is motivation to minimise production costs within the ambits of customer satisfaction so the enterprise can in return enjoy better margin. However, MTS orders on the other hand, are market speculated orders with objective being to meet marketplace manufacturing needs, an order philosophy for replenishment, an inventory type of finished goods and spare parts to support them, a consistent demand pattern, lot sizing based on economic order quantity, and a control concept where all items are equally important, and completion of production released to warehouse to satisfy the demand chain for such finished goods.

Digitalisation of manufacturing Production regiments production using a documented manifest as a production order based on an MPS output when the demand plan is processed on it. The production order being the blueprint of production manifests the item required, the quantity required, as well as date required. Meanwhile these production orders just get outputted from the MPS and need a framework to sequence work around them so as to get them operational, otherwise there is no mechanism to trigger them into production as they are just data. Therefore, order needs to be built around them so they can become working documents of production and in a timeframe suitable for the enterprise to mobilise production resources that gets them safely into work centers and that appropriate work in them to sufficiently achieve the status required to get them out into required packages suitable for the supply chain.

With initial entries completed, the production order absorbs a BOM and a routing attachment to enable it commence its journey into the work centers of production and when an appropriately composed production order is saved the system could estimate the start date and cost of executing the order as a first step based on the attached BOM and routing. When the production order is saved materials and manpower are encumbered for the production order and can be reallocated and reassigned to adjust work execution. More or less manpower could be assigned to further fine-tune production for process optimisation and cost levelling.

The routing sequences attached to the production order provide the length or time required to execute the production order and when calculated against the item determines the incremental cycles of value addition based on the estimated delivery date the order is required to calculate the start date by subtracting the estimated duration from the required date initially entered. If the start date calculated is earlier than current date it means the estimated delivery date is unrealistic and must be moved ahead so the production start date can yield a date in the future for planning. Scheduling then means the specified production order must reach work centers on or before the estimated start date. Also, the routing cost rollup helps to estimate the production cost based on the parametric cost of item quantities included in the production order.

Scheduling arises from an implementation of master production schedule (MPS) where plans are charged up on a time horizon for production of independent materials based on the enterprise demand chain. MPS outputs are then fed into a computerised material requirements plan (MRP) tool where dependent material requisitions are generated with the aid of a BOM input, the MRP system also generates the production orders that will consume machines, manpower, materials, and methods as resources to be managed for production delivery. The MRP system releases an inventory of jobs to be executed in work centers to consume resources towards the fulfillment of material requirements imposed by the MPS for dependent materials. The MPS output is an inventory of production orders for independent items. For each order the due date and quantity of the required products are specified. MPS as a basis for production communication in product terms becomes an input into the MRP tool for BOM explosion to establish gross requirements for production of stated independent items in dependent material terms.

Production Order Queuing To Work Centers

The work center is a basic structural unit of manufacturing activity where machines, manpower, and controlled methods prevail such that when production orders and materials are released to them they all work together to transform the materials into specified items which could be finished products, semi-finished products or sub-assemblies based on the instruction set contained in their production order routing sheets listing the exact sequence of operations required to complete the jobs specified, the route sheets accompany material movements into work areas as well as engagement of manpower to ensure their proper material transformation and handling as well as implementation of production order requirements.

Materials will spend as minimum a time as possible in designated work centers, and being transformed and value added must be transferred to the next work center specified, for its journey of transformation and value addition to be progressively realised.

This resulting list of jobs, being the production orders released from an MPS, are digitally stacked awaiting fulfilment in work centers where they are to be worked through the array of work center cells available in the production facility so that materials can be operated upon by work center machines, manpower, and methods contained in routing sheets to have them transformed into finished products, sub-assemblies, and co-products, but production orders need to be pre-arranged to move through production facilities with maximum efficiency in the time frame specified so as to avoid a costly inventory of work-in-process build-ups which are a nightmare to every manufacturer due to the material handling risks, and cost overrun, that they pose.

When posted to work centers production orders must be checked to ensure they are adequately resourced to deliver. Work center review and coordination helps to achieve load balancing such that resource consumption is evenly spread for graceful throughput of production and moderation of resources for sustainability of the production function.

Every production order queued to a work center must specify a start and finish date based on the content of the attached routing. The determined and specified start and end dates of all scheduled orders help determine the loading of the work center if it is within capacity or above capacity. If queued orders are observed to be above work center capacity then levelling will need to be carried out to smoothen the loading of the work center to stay within operable capacity to avoid production bottleneck or machine breakdown, with production bottleneck too many products compete for throughput putting a strain on manpower and could result in workplace injuries causing lost time, or machine breakdown, either way, bottlenecks are expensive and must be avoided.

Orders get pushed to work centers as long as they are within capacity. The work center will accept these orders and commence production unless it encounters problems, such as overloading, materials gap, skills gap, or machine gap/breakdown in which case it will revert the situation to the planners who will in turn initiate remedial actions and push back the production orders if the remedial actions taken pull through.

Production Order Resource Allocation

Production scheduling encumbers and commits resources to stated works in a manner that does not overburden production facilities so as not to experience consequences of involuntary breakdowns or risk of failing to meet scheduled plans. Then matters of crew management, capacity, and processes must guide and safeguard the schedules to delivery. Planned works from an MPS are not enough, they need to be roster stringed out to absorb available capacity based on available working hours specified by standard hours and overtime hours of factory shifts to perform the schedule required to deliver the finished products.

For production to be properly sequenced for manufacturing effectiveness and functional efficiency work scheduling must be addressed in three stages as stated below:

Stage 1: Production Planning: Planning secures availability of appropriate machineries, work crews of the right manpower, specified materials, and composed production routing methods, appropriately warehoused in work centers in optimal readiness for transformation of materials into the required finished products, sub-assemblies, or co-products. Planning increases the efficiency of manufacturing production to achieve stated objectives and conformance to required quality standards.

The streamlining of production processes ensure that production orders are secured in the most appropriate coordination for manpower utilisation on a workflow that encourages manpower efficiency that delivers greater quality, reducing manpower frustration, and material wastages to effect greater production yield and throughput.

Material requirements, machine hours, manpower hours, and routing sequences all need to be methodically allocated to each production order within stated capacities of the production environment in optimum levels to achieve a best fit cut of capacity utilisation.

A point of coordination need to be provided for workflow control of the repeatable pattern of activities obtainable in manufacturing settings for job based information sharing among crew members to synchronise operational status of production resources to achieve an extension of transformative cycles on materials entering the production mix with adequate workflow coordination such that information regarding how the actions of crew members within the facilities are adequately controlled with job based information sharing between work centers.

Production orders need to be assigned priorities to elevate their visibility for execution when they enter work centers and pile up. This is important as many jobs are released by the MPS and need to be executed and cleared for others to emerge from the MPS mill. Therefore, priority rules are attached to production orders during planning to convey meaning to crew members as jobs approach work centers. The priority rules are: First in first out (FIFO); Early Due Date (EDD); Current Job (CJ); Shorter Processing Time (SPT); Critical Ratio (CR); Slack Time Remaining

(STR); Slack Time Remaining per Operation (STR/OP);

Planning helps to document available capacity so that an estimate of duration for work scheduling can be derived more accurately every time jobs are scheduled to work centers. In the same vein these capacities are further verified by returning actual against planned capacities every time jobs are executed to improve accuracy of future plans.

Allocation of production order to specific work centers will consume machine and manpower capacities in those work centers. The machines transform the materials fed to them while manpower coordinates the operation of the machines.

When efficient planning is engaged an overall level of output is established to derive better yield levels of finished product expectations based on input materials and established capacities. Ultimately, input to output coordination could be managed such that increasing input could increase output or resources could be more exploited to increase output input ratio by innovative scheduling.

Stage 2: Work Routing: Work routing directs the optimal value chain in the conversion cycle that creates the path of significance in material transformation involved in production such that at every work center machine stop there is considerable value addition in material conversion cycles as they are transformed into designated finished products along the production string.

Routings come alive and work within work centers to enable capacities to be disposed towards building value with encumbered work-in-process inventory to enable it leave the work centers opulently for a significant yield to the value chain. Proper routing ensures that jobs are adequately padded up for entry into subsequent work centers that have adequate resources for the transformation expectations stated in their work plan and that they spend minimal time on production lines for best production throughput while bestowing resources to work-in-process to enable them achieve finished product status.

Stage 3: Job Preparation: Job preparation sequences the management of the work of following routings for the production of finished products and moving jobs into the work centers at optimal timings to yield the most economical outcomes to the enterprise by developing sequences to move production orders more economically and efficiently across work centers for the best production yield. The schedules must be adjusted to stay within available inventory, orders, and resources to avoid constricting production into bottlenecks. Scheduling establishes consumption patterns for work center machines and manpower, as well as their timings and costs. Effective scheduling yields higher productivity and efficient manpower engagement.

When jobs are scheduled to work centers where they are to be worked on the following are considered: **queue time**, being the duration the job must wait before being attended to in the work center; **efficiency percentage**, being an estimate of how much yield to expect as production output; **number of machines/manpower**, this is the factor that multiplies the rate at which the routing will enjoy throughput from the work center when applied; **minimum**

scheduled time period, is the duration estimated to be consumed by one full routing roll-up for one item of production; **standard hours**, this gives the number of hours that enjoy normal rate per day outside which overtime is applied.

Work-in-Process inventory builds up while jobs are being expedited and need a systematic focus to prioritise and clear the work stack so that more jobs can come in to be processed and released to warehousing storage without work-in-process inventory creeping in.

Graphical Work Order Scheduling

The Master Production Schedule (MPS) is the main driver of the Material Requirements Plan (MRP) that processes requirements and releases orders for dependent materials to smoothen production of independent materials. Along with the BOMs,
MPS can determine what components are available for manufacturing activities and what components need to be ordered from external vendors in fulfillment of manufacturing requirements priority plans. Aggregate planning creates master production schedules for finished

products. The objective of MRP is to translate those finished product schedules into purchasing and production orders for the entire facility. At a minimum, an MRP system must have an accurate master production schedule, good lead-time estimates, and current inventory records in order to function effectively and produce useful information. The three major inputs of an MRP system are the master production schedule, the product structure records or BOM, and the inventory status records. The demand for end items is scheduled over a number of time periods and recorded on a master production schedule (MPS).

MRP processing first determines gross material requirements, then subtracts from it the inventory on hand and adds back in the safety stock in order to compute the net requirements. The main outputs from MRP include three primary reports and three secondary reports.

The basic functions of an MRP system include: inventory control, bill of material processing, and elementary scheduling. MRP helps organizations to maintain low inventory levels. It is used to plan manufacturing, purchasing and delivering activities.

The purchasing orders are for preferred vendors of materials as stated in the inventory control sourcing records and the production orders are meant to be routed to work centers with relevant resources for their production.

The MPS is an indicator of what, when, and how many items to produce. Besides, the information that the MPS generates can be used as an indicator that shows how much needs be committed to production for the stated output.

Work scheduling with a graphical user interface (GUI) strive to realize smart work scheduling for management of production for data-driven, intelligent, collaborative, and sustainable manufacturing towards knowledge-based production management for efficient production flow planning, demand management, material requirements, material disposition, work scheduling, order processing and production order progress monitoring.

Production Execution

Executing the production plan is what delivers the benefits of established plans for machines, manpower, materials, and methods, as well as defined routings and schedules to warrant that the crew members carry out the described procedures, that resources are available, and that work centers deliver throughput not exceeding their capacities, and efficiently manage the overall processing time of production by constructively reducing work center cycle times to achieve the right quality, the right quantity, right time, and the right manufacturing cost.

Production execution is the final lap of the production process. Having delivered the right plans, routings, BOMs, and schedules for production, execution finalises all these to take finished products to completion, packaged, and released to warehouses with customer satisfaction in its fulfillment, manpower motivation, cost and time reduction, robust order tracking and overall failsafe production strategies. By analysing and optimising production through the various resource matrix the production manager is able to pick the best approach that delivers the most enterprise objective through setting priorities for job management, crew management, work center and machine status monitoring, and in fact the management of all production resource outlay for effectiveness in the delivery of finished products.

Job Management

With planned production orders as the main output of MRP the jobs involved in those orders need to secure free course among the resource outlay of the enterprise for favourable throughput. MRP generated jobs will just sit on the computer as dumb lists that need to be picked up and assigned to work centers and crews embers for execution. The way to get the orders on a path of best progress is to string the generated production orders to available work centers where the machines and methods in them acting under the regulator of available manpower will see to their optimal delivery.

At the work center level priority rules arise to move jobs through established queues so they can be worked on as they arrive the work centers. Priority rules are based on one parameter as simple rules or a combination of parameters for complex rules. No priority rule is best for any given work center, simple priority rules are as follows:

1. **First in First Out (FIFO)** – this priority rule is based on the arrival order of jobs. Jobs are processed on each machine from the previous operation. This rule is very simple to implement.

2. **Early Due Date (EDD)** – the jobs are scheduled according to their due date. The earliest the due date the higher the priority of the job.

3. **Current Job (CJ)** – this priority rule is designed to save set up time by processing all jobs that require the same set up in sequence, thus eliminating the need for set up operations. The problem with this rule is that jobs with an early due date may be delayed while other less urgent jobs are being processed.

4. **Shortest Processing Time (SPT)** – this priority rule tries to minimise the number of jobs waiting in front of a machine, by processing short jobs first. The problem with this rule is that long jobs may be completed late while short jobs are completed earlier than their due date.

In addition to the simple priority rules there is a variety of more complex rules including the following:

1. **Critical Ratio (CR)** – this rule is based on the difference between the due date and the current date divided by the time required to complete the remaining work. Jobs with smaller value of CR get higher priority.

2. **Slack Time Remaining (STR)** – this rule is based on the difference between the time remaining before the due date and the time required for processing the remaining jobs. The smaller the value of STR the higher the priority of the job.

3. **Slack Time Remaining per Operation (STR/OP)** – this rule is based on the average slack time per remaining operation calculated as the

ratio between STR and the number of remaining operations. Higher priority is assigned to jobs with lower value of STR/OP.

These priority rules are expected to be assigned to production orders as priority constraints to guide the orders through work centers to secure for them a heightened throughput for greater yield such that the most suitable priority rule based heuristic is chosen to deal with resource constrained production execution. When production orders arrive a production queue in work centers their priority of work are determined chronologically by determining the arrival dates of the production orders, thereafter priority rules help to place execution focus on what matters.

Crew Management

Production manpower must be well utilised. As production talent, the crew learns a lot on the job everyday and as such must readily assimilate learning experiences for enterprise success. Team knowledge makes sure that every crewmember knows their default work routines and is ready to take on and deliver every new assignment assumed. By working a crew list and calendar the crew available is of significance to the execution of production orders as the team are sometimes optionally available, on leave, or on training. Management of crewmembers must work towards less of sick leave and injury time by making the workplace safe through observance of safety protocols and safe handling of materials. Environmental safety is not only for material value retention but also to safeguard the health and well being of personnel at work, thereby securing future throughput.

When workplace inefficiencies are surveyed and worker safety enhanced there is reasonable ground to suggest that machine efficiency will be sustained to assure hazard free workplace. Access to delicate machines then must be limited only to those manpower that have been properly guided by introduction of access control to limit access and guard against machine operation by otherwise inexperienced or unsafe hands. When machine failure rates are constructively curtailed more capacity will be made available for production and performance targets can be guaranteed.

Management of Production Resources

Effective execution of production operations to maximise yield through the technical cross-checking of resource readiness and availability frees resotces for next production orders on queue.

Next is monitoring of machines and work centers for successful implementation of production orders and an understanding of how conditions can be optimised to improve production output, ultimately tracking and documenting the transformation of raw materials to finished products to improve yield and processing time to align manufacturing expectations with resource capacities and availability.

Driving the manufacturing production for best yield possibilities is based on assessment of the driving factors for productivity as key to increasing sustained finished products yield, otherwise there could be significant reduction in production output. By improving training, quantifying everything, organising everything, standardising manpower, implementing cellular production, proactively managing equipment uptimes, and strengthening enterprise supply base the production manager can place manufacturing production on high efficiency and drastically reduce wastage and spoilage of production runs.

Despatching produced finished products and co-products is a significant output of every production execution system for on hand inventory increments and of every conversion cycle to account for manufacturing costs by way of back flushing materials and overhead. Production output must essentially be delivered in quality packaging to sustain value retention beyond the demand chain's revenue cycle to account for duration of trucking to shelving positions.

Performance analysis is important for manufacturing after allocation of raw materials, manpower, and work centers to fulfill production orders. This operation at this stage involves systematic observations to enhance performance and improve decision-making by studying and evaluating the performance of particular scenarios in comparison to expectations of objectives to be achieved.

Quality management when its time to release finished products from manufacturing, production execution must ensure that quality products are released based on acceptable quality, appropriate quality, and aspirational quality with assurance, planning, and control with appropriate actions and test procedures for effective post production marketplace perception management.

Management of production activities through efficient loading of production orders to work centers and sequencing established routings and methods enable adequate throughput to consume plant capacities sustainably and release finished goods of the right quantity to inventory at the right time.

Management of production capacities - production targets, key performance indicators (KPIs), production by MTO or MTS require different levels of capacities, conversion cycle efficiency maximises capacity, production parameters, appropriate skillsets, division and specialisation of labour, proper use of tools and machines, setting up facility is cost prohibitive so production must engage close to fully installed capacity.

Management of production yield is essential for high volume manufacturing as consumption ensured even with identical nature of the outputs depletes more inventory as defined by the attached BOM for profitable production requiring fewer manpower and engaging more machines and power tools resulting in lower per unit cost. One-off, batch, or mass-production benefits from engaging machines and skilled manpower with greater product knowledge. In the end the yield from production execution determines how the balance of time, money, and quality as important factors has benefited the enterprise.

Production Release Back-Flushing

Production processes need proper accounting to show back events hitting the various cost centers connected to production in a graceful pattern to ensure the books are adequately posted and that accounting offers a proper language to the events going on the shop floor in a way that cascades to the final accounts of income statements and balance sheets. The way this is done in manufacturing production is by first loading materials, labour, and overheads into work in process (WIP) accounts with journal entries to the WIP journal to initiate production for which the WIP journal posting is performed by debiting the relevant work in process account and as required by double entry crediting the designated accounts of materials, labour, and overhead with a sum totalling the same amount posted to the WIP account so that the debit equals the credits inputted. The WIP stack elimination is a true representation of full back-flushing, but partial elimination of the WIP stack is a reflection of partial back-flushing due to lower yield and wastages as a result of material shortages limiting production and capable to result in a WIP variance report.

When back-flushed, materials, labour, and overheads are stated as consumed and their respective costs are encapsulated or absorbed in the cost of the finished products posted. Back-flushed quantities could either be based on the quantity of finished products stated on the production order or the quantity received from production at closeout of production.

At completion of the production order, a journal entry is made to the production journal indicating the end of production. When this production journal is posted the materials, labour, and overhead are considered back-flushed. The back flushing is indicated in the manner the accounting is conducted, the accounts are posted in this case as a reversal of what was earlier posted at initiation of the production by debiting the finished products inventory by how much products have been produced to increase the on-hand quantities of inventory and crediting the WIP account by the value of the finished products inventory that was posted, signifying that the materials, labour, and overheads have been transformed into finished products.

Capacity Management

In this era of social change and technological progress, enterprises that see the big picture demand marketplace domination and service greater needs by producing more for the market they control. Capacity management is key to production execution, as production managers must schedule production within available capacity of production resources at their disposal for a sustainable production engagement. There are consequences for operating at optimum capacity such as equipment breakdown and required engagement of maintenance procedures rather than commercially beneficial production. Costs placed in maintenance are sunk costs and may yield further liabilities and expenses from the expenditure cycle if the maintenance undertaking is not successfully carried out.

Capacity is the maximum level of production that an enterprise can sustain to make finished products. Enterprises operating within their factory capacity nameplate can plan to raise their future capacities to meet greater demands. This means limitations of the production process must be accepted and planned into production and that current MPS output must be limited to available capacity, except long term MPS output that could be budgeted for constructive capacity increase.

No manufacturer has infinite capacity. The available capacity must be considered separately as human and machine capacities. Human capacity expressed in summation of hours is a product of actual working hours, attendance rate, direct labour rate and equivalent manpower. While machine capacity expressed in hours summation is a product of operating hours, operating rate, and the number of machines.

Every production order requires capacity to deliver without which it will be perpetually queued. Managing capacity moves the enterprise into the realm of having machine nameplates at your fingertips for the purpose of safeguarding flourishing production timelines.

Of competitive advantage to the enterprise is the concept of capacity planning as capacity of production resources determine how much can be taken without compromising production resources into a bottleneck due to low production throughput. Therefore enterprises must take great precaution to have an industrial scale installed capacity to address production needs and even make long term plans with budgetary inputs to address long term production requirements based on long term demand forecasts as a competitive strategy.

Work Center Load Management

Pushing the envelope of production resources by building more capacity for uptake by industrial customers deploys production resources to take on more orders at reduced loading and lower failure rate.

Capacity utilization rate being the extent to which enterprise capacity is employed towards production significantly determines how valuable the value chain to which the enterprise is connected is beneficial, utilising more capacity is expected to increase the ratio of output margin to input materials.

Production status management is paramount to executing production. Production moves from order initiation (OI), as a work center accepts the production order. Next the status of the production order moves to order planning (OP), where the work center management accumulates materials to execute the order and assigns manpower of requisite skills. Subsequently the production order moves into execution status (OE), where work center manpower tears down the production routing attached to the production order transforms encumbered materials, runs the machines to transform materials as specified in the routing sheets and assignment of tasks and workload balance to ensure production is adequately carried out such that production control will subsequently be invoked to galvanise the various moving parts of the assembly line for supervision and control.

Eventually, packaging secures that the encasing of finished products in a primary packaging for customer handling through the demand chain fulfilment strategies adequately retains value for flourishing consumption of the product through its life cycle. Beyond the primary package is the secondary package to aid product handling by distributors and retailers as stakeholders of the value chain where the product thrives. The secondary package sustains distributors and retailers investment in the distribution of the product through beneficial packaging that takes their interests into consideration particularly for product vending and dispensing. A step above distribution is the need for bulk movement of products through a somewhat more compartmentalisation of the product for pallet encasement through the distribution chain's fork lift systems, trolleys, and conveyor belts for forward logistics through distribution centers for controlling bulk movements of the product to the final consumers economically through tertiary packaging that encapsulates multiple secondarily packed units into a considerably more protective package that secures the product from handling concerns as it moves through distribution channels.

Warehousing is a value retention infrastructure that acts as landmark of the distribution channel. Warehouses holds finished products to a stop and ensure that their values are securely kept away from all forms of corruption.

Production release must be achieved in the long run in spite of not having 100% resource and capacity outlay. Production must be rolled to the finish line even with dwindling resources and capacities at the disposal of the order pool. By deftness in knowing what need to be augmented so that warehouse release can be concluded for WIP accounts to be back flushed may require a little more manpower time and somewhat above machine capacities to safeguard work center throughput.

Goods receipt against production orders signify the end of production as they are handed over to warehousing to be held in storage. These could be bulk goods or batch goods that have been completed in line with a production order. Releases of the goods mean a notification to the demand chain that there is readiness for fulfilment of delivery that will be arriving soon.

Where to place material disposition as they arrive the warehouse is in the incoming goods staging are so that all documentation about arrival to warehouse could be completed so that storage location and inventory control could be updated to reflect the warehouse holdings for the day.

Optimization of End-to-End Manufacturing Processes is what

scalable management of production resources means. From modeling to machining there is a lot that needs to be optimised before the final product output to warehouse. Most importantly the product specification formulated in the BOM and the production methods composed in routing statements are the most significant manufacturing data that must be constantly updated with innovations around them for the best of timing and cost control.

Enterprises seeking to enjoy competitive advantage must optimize their end-to-end manufacturing processes for improved visibility.

Production Planning

Efficient planning overcomes production challenges. Capacity required to blueprint, structure, and scheme production resides with the white collar that sits on the planning table to make sense out of the numerous production orders generated from the MPS to be moved across shop floors and work centers to consume production resources while yielding reasonable product counts in the process.

Modern tools that aid production planning attempt to provide visual tools to envision what needs to be done to move production orders through desired work centers. The Planner's job is made approachable with digital handles that place control on production order progression, ensuring that every production order passes through order initiation (OI), order planning (OP), order execution (OE), quality control (QC), and order release (OR). The planning that guarantees successful order execution must ensure that bottlenecks are cleared early before each production order is attended to.

Production planning function is generically subordinated to the Production Resource Manager who is the production resource line manager for the enterprise. The Production Resource Manager coordinates the work of Material Planners, Routing Planners, Production Managers, and the Planning Managers. The Production Resource Manager (PRM) ensures effectiveness and efficiency of infrastructures of logistics equipment of the BOM formulation process, production work centers of the routing composition process, planning boards of the work scheduling process, and shop floors of the production execution process on which production processes and mechanisms choreograph the enterprise production strategy.

Operations that Planning must get right to deepen the planning function include inventory management, procurements, maintenance management, project management, capacity planning, contracting, sales ordering, as well as BOM formulation, routing composition, work scheduling, and production execution which are direct processes for managing production resources.

In production planning the Master Production Schedule (MPS) is a major manufacturing resource that lays out the production strategy that rolls up the BOM and inventory status into an executable manufacturing plan spread out across work center timelines.

MRP alone results in over and under shop loading but must be augmented with capacity requirements to balance out work center capacities and demand plans.

The risk and reward balance that MRP and MRP II brings to the manufacturing planning table through MPS, BOM, and capacity plan integration results in good product options and favourable pricing strategies with commendable market intelligence. With MRP having grown to include capacity planning, shop floor control, and procurement by synchronising materials with production requirements for a modern and data driven MRP II system.

The MPS delivers the production program on which manufacturing delivers the product quantities required and at the right time. More important is the simulation and scenario planning capability of the MRP II system. With integration being a key ERP issue, the MPS is a major resource that releases integration handles for all enterprise relationships and resources such as CRM that takes in customer orders, the FRM that accounts for material and activity costs, the PRM that programs the execution of production, the TRM that manages the talent capacities required to man relevant work centers, the VRM that handshakes procurements with vendors, and the WRM that holds required storage capacities for input materials and finished product output. If these systems offer full integration then that must be a state-of-the-art as ERP requires full integration to avoid frequent rework and patching of manufacturing data entry activities but is an advocate of a write once use many times approach.

With MPS being central to the manufacturing enterprise, there is reasonable virtualisation of capacity for work centers and crewmembers to produce in accordance with industry specifications based on customer orders and material procurement from virtual vendors to address the shift away from make-to-stock type manufacturing to a modern make-to-order scenario that sits well with a modern digital oriented enterprise.

The concepts of MPS, MRP, MRPII, and capacity planning are deeply rooted in manufacturing and as such provide the keys to all aspects of enterprise integration and digital transformation.

Planners and schedulers hang out in factories to ensure adequate content for the MPS that sequences production resources and that they are free of bottlenecks that often becloud the planning and scheduling function when multiple products are involved, while at the same time delving into the decision support system that enables the planners and schedulers resolve the chaos confronting them daily. The human factor is still heavily involved in modern production planning despite scenario options available with AI in modern planning and scheduling software. The human factor still holds the ace in modern manufacturing since planning must be top notch to deliver scalable results daily.

Having laid out what we want to do with planning and scheduling of production resources the first question we want to ask is, what is Scheduling? And, What do we mean by Scheduling? The goal of an efficient production resource management system is to achieve a healthy manufacturing environment for the use of best practice production control so that actual production can follow developed schedules created from crafted plans and coordinated resources that ultimately improves the effectiveness and efficiency of manufacturing plants.

Planning, Scheduling, & Dispatching

Production planners have to make room for the activities of planning, scheduling, and dispatching to happen in the manufacturing enterprise to adequately capture the essence of production. These are carried for different goals and phases of manufacturing. Planning is carried out at a higher level to capture the absorption of customer orders and relate them to production output so as to realise production goals with appropriate resource arrangement over a fairly long or medium term time horizon involving the whole enterprise. Scheduling focuses on short-term production execution to realise material consumption, management of production resources, and precise deployment of production crews to work centers for capacity management. Whereas, dispatching is real time production activity for the physical release of production output quickly into warehouse storage systems for value retention immediately they are produced.

The specifics of these phases require precision tools in all cases to recognise and manage value creation by the enterprise so as to elevate manufacturing yield and reduce scraps to the barest minimum. Besides, it is important that KPIs be driven off these tools for the enterprise dashboard so as to make manufacturing a goal-oriented endeavour in value creation for the enterprise so that practitioners are kept focused on the big picture and always on their toes.

Master Production Scheduling

The key tool for planning manufacturing is the Master Production Schedule (MPS) that lays out on a time horizon a master plan for what needs to be produced, when, and how many to be produced based on prevailing supply and demand scenarios. The MPS is the backbone of manufacturing enterprises and as the main driver of production it captures and rolls up expectations in all constituents of the enterprise that supports production such as sales, marketing, engineering, technical support, and finance. The production-oriented MPS has proven a most effective tool worldwide for laying out manufacturing programs; its effectiveness stems from the fact that it has the mechanism to absorb what is required to be produced as input variables and lays out what needs to be done to achieve what is required as its result set. As an essential tool, the MPS provides the master plan of production indicating the quantities and time periods known as time buckets at which they are to be produced.

When firmed up, the MPS sends its output to the MRP so that BOM explosions can give dependent item material count requirements to produce the independent item product counts specified by the MPS. The dispatching specifies how the manufacturing outputs are accumulated in warehouse storage for value retention and customer fulfilment towards achieving favourable available to promise (ATP) numbers.

The MPS works both the demand side and supply side of things. The demand side brings in customer orders for fulfilment operations while the supply side routes purchase orders for material requirements to vendors for procurements.

Planning with an MPS will subsequently generate estimates for manpower, machine, and material requirements that need to be in place to deliver the end products specified to be delivered from the facility. The attached routing sequence supports the generation of machine and manpower requirements based on the manufacturing methods contained in the sequence tables of routing statements, while the attached BOM when juxtaposed with the product counts generate the material requirements which of course are initially gross requirements and subsequently phased through netting and offsetting protocols of the MRP tool to average out and derive the appropriate interface to the supply chain to land materials in the production facility to meet material gaps just in time. If vendors are hot-linked to the enterprise inventory and validated for Virtual Requisition (VR) of netted orders offset to reach the enterprise on time, then orders will be electronically projected and dispatched to vendors for expediting and when authorised the vendors release Advance Shipment Notification (ASN) with tracking to rush materials for production to key in the supply chain's participation with considerable reduction in lead time.

MPS is a simple yet very important enabler of manufacturing. The range of virtualisation that the world earnestly awaits now is digital virtualisation to deliver supply chains that sense and respond virtually to the physical world, being a supply chain for models that are searchable across the internet with precision. Therefore, the right MPS model the world needs should quickly put state-of-the-art scheduling into the hands of many users virtually in real-time with no local computer, local area network, and no server other than browsing to the internet with data and site subscriptions. Such state-of-the-art scheduling should get SMEs to possessing the virtual cutting edge of technology with capacity to enable robust scheduling and running the sustainable production of everyday consumables and consumer goods that the world needs to a step into the virtual realms of production, processing, and packaging.

With the MPS being a versatile manufacturing tool that works with inputs, controls, mechanisms, and outputs, there is need to have performance measures tied to the operator schedulers for effectiveness of the manufacturing enterprise. What is required is for monitoring of the MPS to promote best practice in production and effectiveness in consumption of production resources. The MPS therefore must deliver high productivity for manufacturing facilities with precision scheduling, focused production, and seemly dispatching.

Being an aggregate planning tool, that comprehensively outlines the scheme of units for producing end items, the MPS firms up the enterprise time horizons in specified time buckets, freezing some time zones known as time fences from accepting more orders from the demand chain so as to focus on production of accepted orders and other time zones in future, known as planning time fences, for accepting and planning orders for MTO production that are based on actual demand from customer orders, while MTS production have regularly controlled inputs into the MTS from demand forecasts. The yield of the MTS is expected to provide decent available to promise (ATP) units that feed the demand chain.

Planning Lower BOM Levels

Lower level dependent materials required to produce end items are specified and structured in BOMS to formulate product structures. The planning that gives rise to their consumption in manufacturing facilities absorbs the BOM into an MRP system that yields BOM explosion that gives material counts required to produce what is planned as the MPS output, requiring the multiplication of the BOM quantities by the planned counts of the MPS output for independent materials. This disaggregation of planning is done not only to explode the BOMs but also to net out requirements against on hand quantities to save cost as well as offset the timing or vendor orders so that requirements are in stock just in time to effect production and realise favourable available to promise (ATP) scenarios for end products.

MPS & BOM Input To MRP

Away from the sizing and timing of production objectives of each independent item with the MPS aggregate plans. With the MPS on hand to generate independent item counts required to be produced the complimentary MRP system takes the associated BOMs of the independent items specified by the MPS as inputs into the MRP system to generate the **BOM Explosion** process that provides the material gross requirements as a disaggregation of materials to go into production as a product of material quantities and end item counts to go into a sequence of machining and assemblies in work centers by essential manpower to deliver the required finished products by operating the supplementary production routings composed for the specified end items.

Next, the MRP does **Netting**; a process where comparison is drawn between MRP derived gross requirements and on-hand inventory of dependent items so as to subtract what is on hand from the gross requirements to establish net requirements for production. The net requirement is what gets ordered from vendors, substantially reducing cost of ordering gross material requirements instead.

Finally, the MRP process provides **Offsetting**; here, the right timing to initiate vendor orders are generated for the net requirements established and taking into account the vendor's lead time so that the materials are just on hand at the right time required to effect production minimising storage requirements and securing vendor commitments to production.

When it comes to capacity plans both MPS and MRP have their different capacity plans that they generate. The MPS outputs Rough Cut Capacity Plans (RCCP) while the MRP outputs Capacity Requirement Plans (CRP).

RCCP and CRP affect production in different ways. RCCP is achieved at the beginning stages of MPS run whereas CRP follows later for capacity check for the MPS. RCCP is based on what is to be produced but CRP is based on what is ordered from vendors. RCCP gives rapid feedback whereas CRP gives feedback slowly.

MPS Performance Measurements

The MPS has grown from being just an MRP input into a management function for production coordination. MPS being a management function demands efficient and effective implementation with control handles to elevate performance functions and establish greater control of the process and mechanism of production.

It takes much to oversee production and manage a production facility; decisions must be made to achieve enterprise goals. Key Performance Indicators are promising when it comes to measurability of manufacturing performance. Manufacturing KPIs provide data for measuring manufacturing performance, but what exactly do they represent?

A key Performance Indicator is a measurable value that indicates how a facility is effective at achieving its objectives. They are a type of measurement that helps manufacturing functions understand how the facility is performing. Manufacturing enterprises use KPIs to analyse, monitor, and enhance their operations. Each KPI need to be linked to a target value so that they can be evaluated as meeting expectations or not.

KPIs abound for efficiency, effectiveness and improvement of MPS operations for best practice. Efficiency KPIs save resources, while effectiveness KPIs deliver MPS service improvements for better scheduler productivity.

Benefits abound from implementation of manufacturing KPIs, such as: accountability, visibility, decision support, profitability, performance enhancement, and waste reduction. Top on establishing an effective management profile for an MPS is the idea of providing an MPS KPI capable of dash-boarding metrics of the range of numbers attributable to definite measures of performance to indicate how effective the planning function is coordinated to deliver best in class manufacturing metrics. Manufacturing KPIs will be deployed to determine the course of action to be taken on the following five challenges:

1. Monitor enterprise health
2. Measure production progress
3. Make adjustments and stay on track of production
4. Solve production problems and tackle opportunities, and
5. Analyze production patterns over time

Measurements are the feedback loop to both performance and improvement in general. The master production schedule (MPS) is one of the most important processes in an enterprise resource planning (ERP) business system and therefore should have robust measurements to drive that performance

Manufacturing KPIs are for measuring manufacturing schedule adherence and also for setting measurable goals for manufacturing facilities. The indicator for measuring actual production as a percentage of scheduled production, also known as the absolute variance. This KPI shows the variance between the two figures of actual and scheduled production.

The formula for this key metric is:

$$Absolute\ Variance = \frac{\#\ of\ Work\ Orders\ Delivered\ On\ Schedule}{\#\ of\ Scheduled\ Work\ Orders\ Due}$$

$$Schedule\ Adherence = \frac{Scheduled\ Production - Actual\ Production}{Scheduled\ Production} \times 100\%$$

In addition, production need to have KPIs for such things as capacity utilisation, cycle time, equipment effectiveness, throughput, yield, downtime to operating time ratio, scrap rate, inventory turnover, availability, and maintenance rate. All these are better available on a production system dashboard for effective tracking and reporting of production activities and resource utilisation.

Production Management

Production management is an activity that organizes, directs, controls and enables the production function to gainfully yield finished products in a manufacturing environment. Production management's role is to help the enterprise achieve its business objectives of producing the finished products required to satisfy customer demands. If customer demands are met sales targets will be realized and revenue cycles will kick in, to further reinvigorate the conversion cycle that turns the transformative manufacturing value addition processes and mechanisms and the expenditure cycle that drives the supply chain keeping the enterprise flourished.

The production capacity of the enterprise then becomes defined by the combined effect of how the mechanisms of the conversion cycle interchanges the revenue and the expenditure cycles to derive a favourable margin as well as the effectiveness and efficiency of production that the production management is able to achieve.

Management of production aims to coordinate various components that enter the conversion cycle's value addition process through 'the five Ms' of production, being machines, manpower, margins, materials, and methods to consistently derive a measurable margin for the enterprise.

If replenishment cycles are satisfied and material requirements are met the production function will be in readiness to deliver the finished products required to satisfy customer demands and robustly respond to the supply chain.

The production resource manager of a production function must therefore ensure that his team of material planners, production planners, schedulers, and production managers are effectively playing their roles so that together all production resources right from the work centers efficiently deliver the most finished products to the satisfaction of customers of the enterprise and secures a marketplace competitive advantage in the by so doing.

Machines transform materials that they are fed from one form to another. Production requires a varied set of machines that consume energy to leverage manpower in the transformative process that converts materials from one form to another. For machines to be up and running several factors must be considered and guidelines followed, such as not exceeding their stated capacities and that they must be operated by skilled manpower that understand their proper working boundaries. Also, when they encounter prolonged engagement in operation they could breakdown and must be subjected to a maintenance cycle that consumes a set of materials known as spare parts in conjunction with tools to be used for corrective maintenance so that they can bounce back into optimal capacity. Therefore, machines must be inspected at intervals to determine their readiness for production operations and in certain cases of possible wear-out could be scheduled for preventive maintenance so they do not become a bottleneck in the production process in the event of a break down at a critical time of scheduled production.

Machine Management

Machine management require proper coordination to secure reasonable up time and run time to take on material transformations into finished products. This requires that manpower be skilled in the proper handling of these machines in the work centers where they are installed and that the machine nameplate specifications are not exceeded to avoid breakdowns. For industrial settings there are always industrial grade machines purpose built for industrial use rather than consumer grade equipment that are not capable of industrial payload and duty cycle.

This differentiation of industrial and consumer grade equipment must be religiously adhered to when deploying machinery for industrial operations to avoid production bottlenecks due to machine breakdowns.

In industrial settings, machines are clustered in work centers with skilled crewmembers as manpower for effectiveness. Machine pools could further be streamlined into production lines for effective production throughput and flow control. Machine uptime is guaranteed by roles of maintenance supervisors who carry out machine inspections and schedule maintenance operations when required for corrective, preventive, and predictive maintenance operations. Machines clustered in work centers are under the control of work crews that receive production orders scheduled to them for production and are guided by laid down production sequence that must be followed and rolled up to achieve enterprise production goals. A well written production sequence, often called a routing sheet specifies the activities or operations to be performed using a work center's machines and how the work is to be performed as well as time ceiling to be spent doing the job for productivity management.

Equipment maintenance function must be put in place to ensure ready availability of equipment at all times that production is scheduled on them.

Manpower Management

For effective and efficient management of production resources industrial manpower must be deployed to operate machines and interpret production sequences composed in production routing sheets.

Industrial manpower are considered as skilled manpower because they undergo considerable training to be sufficiently talented for industrial facilities where they are graded according to their skills, knowledge, and abilities. Their skills are a result of their exposure to certain operational routines that have engaged their faculties thoroughly. Their knowledge defines the academic pursuits they have engaged in educational institutions where they have acquired certain qualifications. The abilities of manpower are a result of engaging their faculties through skills and knowledge for outstanding performance.

What manpower brings to the enterprise is capacity encapsulated severally in the skills, knowledge, and abilities that they posses. Factory manpower is usually grouped in crews that are assigned to specific work centers where they work as a team to achieve specific production goals. In such cases they are considered as specialized manpower and their jobs and responsibilities are designated and structured to achieve division of labour.

With the world moving into a structured handling of competence based manpower rather than task based, it is now the talent composition that are of significance as every enterprise thrives to build a talent gallery composed of enviable competencies to handle its resources and achieve favourable results in its relationships with necessary customer and vendor entities.

Margin Management

Prior to embarking on production, management sets out a margin target through a process considered as Value Chain Analysis where costs are ascertained to estimate a margin proposal for the enterprise when the product eventually reaches the consumer. So costs will have been frozen prior to engaging production. Therefore an efficient management of production is that which optimizes production without cost overrun by keeping costs of materials, manpower, and overhead below ceiling as determined from prior Value Chain Analysis. This ensures that the enterprise realizes a reasonable margin from its production activities.

Modern production has greater control over activities in its quest to derive profitable margins by adopting future proof accounting models such as activity based costing (ABC) that tracks costs rather than just allocating them to overheads for best in class money management strategies in the enterprise. With ABC the costs of production can be more accurately tracked and traced leading to greater efficiency and effectiveness in the pricing of products landing the marketplace giving the enterprise a competitive advantage.

Measuring value in production as values are added and converted from one form to another ensures the enterprise earns enough margin values through the stages and cycles of production.

Routing statements are the elements that trigger the cost drivers into cost pools of production cost management as every routing statement executed encumber costs that must be accounted for. Materials are consumed, machine time used, time allotted takes on factory overheads as well as piecework rates to be paid out to manpower crew engaged in production. Materials are issued to factory work centers from warehouses and accounted for by debiting work in process asset account and crediting assets of materials, labour, and factory overhead with the same amount to initiate production. Therefore, when production is concluded a back flush process is initiated that debits materials, labour, and overhead accounts, thereby wiping out values that were earlier credited at initiation of production.

Accounting for materials follow an intricate process of individual material valuation but due to inexpensiveness of technology every material movement can be accounted for in a process considered as perpetual inventory accounting where every material movement is accounted for in a somewhat self balancing double entry inventory accounting system where all material debit and credit transactions draw out a running balance to individual inventory accounts of materials, from material cost layer records, whether based on Last in First Out (LIFO), First in First Out (FIFO), or average costing. These are reflected in inventory control process of the Warehouse Resource Management (WRM) system.

Costing also has to be made for piecework rates where manpower talents are compensated for piece rate production as motivation for producing more. This is based on production targets set at productivity bargaining meetings with manpower crewmembers and specified in the payroll process of the Talent Relationship Management (TRM) module so that these earnings can be reflected on employee payslip.

In ABC costing that tracks and traces costs the enterprise enjoys the benefits of defining cost drivers as units on which costs are based, cost pools act as cost centers that take the drivers of the activities defined based on enterprise cost parameters. Most importantly, enterprise accounts are rendered from the four cycles of ledger, revenue, expenditure, and conversion cycles. Further, there are two traditional types of cost accounting systems, Job Costing and Process Costing.

Job Costing estimates Cost of Goods Sold (COGS) from a recognition of the cost of completed jobs that are based on credit balance of Work In Process Inventory debit balances entered for direct materials, direct labour, and overhead applied when production was initiated. In job costing, completed jobs and their unit costs are recognised. The accounts maintained for job costing are (1) Materials Inventory account, (2) Labour account, (3) Manufacturing Overhead, (4) Work In Process Inventory, (5) Finished Goods Inventory, and (6) Cost of Goods Sold. These 6 accounts help to track materials purchase, trace incurred factory labour, and accumulate incurred manufacturing overhead while being able to assign material consumption, factory labour used, overhead applied, recognise finished products completed, and cost of goods sold.

Process Costing is used in manufacturing processes that have to do with continuous production, known as mass production, where the identity of direct costs of materials, labour, and overhead are considered substantially lost due to simplification of production and cost lowering. Process costing essentially keeps track of what is produced and at what cost. The direct costs of materials, labour, and overhead in process costing are posted to departments or cost centers and the individual department costs are applied on the number units of finished products produced. Mass production moves through assembly lines where they are worked on and moved to the next work center for subsequent work until finished. Such cost simplification as lumped department by department in mass production or process costing helps to keep costs suppressed due to a considerably larger volume of production in mass production which when averaged down to unit cost is substantially less, ultimately resulting in higher margin for the enterprise.

Material Management

Production transforms materials into finished products. Materials for production are often formulated in a bill of material (BOM). Within the BOM the formulation is what is required to produce one unit of the finished product, being the required measure in the unit quantity known to convey the material requirements for the finished product.

Materials formulated in BOMs are expected to be held in inventory prior to being issued to factory floors and work centers for transformation into finished products. As fast moving items their availability must be secured through strategic alliance with preferred vendors to minimize production downtime or factory wait occasioned by non-availability of critical materials to scheduled production. Stocking of BOM materials must follow the guidelines of economic order quantities (EOQ) so that safety stock levels are maintained when orders are placed with vendors for their replenishment at specified reorder quantities which when they arrive are able to move stock levels to the prescribed maximum stock level for the materials while the cycle moves in it's understood pattern and consumption moves at previously specified rate material reorder will be initiated with the preferred vendor whenever the reorder level is triggered again.

In today's virtual world, the reorder level should not only initiate an internal requisition process but should virtually handshake the supply chain for Virtual Requisition (VR). VR is a phenomenon that can elevate the status of the supply chain as a virtual stakeholder of every conversion cycle as that critical agent that brings in required materials with little handshaking whenever they reach critical levels that require steady replenishments and that indeed can keep the world running.

The major material management process of the enterprise is with material requirements plan (MRP) where the product structure of materials required for the production of finished products is derived as required in production. The MRP derives input from two important processes being the master production schedule (MPS) and the bill of materials (BOM). The MPS is the aggregate plan that outlines the production of specific quantities of finished products on a time horizon while the BOM is the product structure of each type of finished product to be produced, showing the quantity of item components that makes up a unit of the finished product. Managing the MPS > BOM > MRP mechanism of the enterprise leads to very accurate material management that readily responds to market situations and gives an enterprise the competitive edge in the marketplace.

Method Management

Guidelines for production activities are encapsulated in routing statements composed to regiment competent interpretation by skilled manpower in work centers as working systems and implemented under the transformative application of industrial machines. Initiations of production methods are commenced from discharge of production orders to work centers. These work centers have the manpower, tools, machines, materials, and environment to tear down the production orders, receive materials and sequence them through various machines for transformation so that the input to work centers being raw materials are transformed into work in progress as they move from work center to work center while on a production queue following precise directives in routing statements. The transformation that the material experiences is often documented in a routing sequence that specifies the type of activity to be performed on the material, the tools to be used, the machines to be applied, as well as the knowledge, skill, and abilities of manpower talents to run the work centers.

The routing sequence of method management helps to determine how much actual time and cost are involved in the transformation of raw materials through work centers as these are achieved by filling out a few entries on the routing sheet as they leave work centers. The routing steps as documented by the production planner at some point need to be associated with BOM items to initiate material consumption when the routing step is executed.

Coordination of Planners

The Production Resource Manager coordinates the activities of the Material Planner and Production Planner. The Materials Planner oversees inventory of materials required in production to the effect that BOM materials enjoy ready encumbrance and availability to work centers for transformation into finished products. The Material Planner documents and revises the BOM for every finished product undertaken by the enterprise in the most economic way inclusive of alternative materials possibly on each BOM item. The Material Planner is empowered to get involved in material sourcing and procurement meetings for enterprise knowledge of BOM content availability and understanding of the criticality of materials at all times for the smooth running of enterprise production activities. Further, if different classes of a finished product are to be moved into the marketplace, such as premium, mid market, and low budget variants, it is the material planner's role to devise a material mix to the enterprise competitive advantage in the marketplace.

The Production Planner documents the operation sequences of activities, methods, machines, tools, and timings required to transform materials into finished goods in work centers equipped with dedicated machines and skilled labour. The Production Planner creates routing sheets with routing steps. It is with the routing steps that materials are transformed from one stage to the other as they move through work centers.

The Production Resource Manager coordinates both planners by getting them to link their outputs of BOMs and routings. The routing steps are to be associated with BOM items to convert them into production routings. Routing step to BOM item association creates a consumption pattern for materials in the work centers to the effect that when finished products are considered to have been released to warehouse an operation can be triggered to back flush all established BOM items considered linked to the associated production routings while at the same time recognising the associated machine costs to be recorded by the cost accountant in the related conversion cycle of the financial resource management module.

Efficient coordination of the Material Planner and Production Planner results in accurate BOM for the management of production materials and effective production routings ensuring that production resources yield beneficial value chain analysis for the best of finished products able to command their own in the marketplace. The work of the planners enables the scheduler to create BOM and production routing attached production orders and easier scheduling to work centers where the production is carried out. In particular the material planner helps to develop a good material requirements plan (MRP) in conjunction with the scheduler who produces the master production schedule (MPS) based on finished products to be produced as input for the production of the MRP.

Coordination of Scheduler

The Scheduler is the mover of the enterprise manufacturing function. As customer orders drop in the scheduler responds with production orders to accommodate them. Customer orders are responded to by the Scheduler as make-to-order (MTO) production orders while warehouse replenishment orders are made into make-to-stock (MTS) production orders with stated production order quantities based on the customer order quantities and established reorder quantities respectively.

Scheduler coordination helps to achieve a more accurate master production schedule (MPS) as well as a good material requirements plan (MRP). The MPS is an aggregation of finished product counts for production while the MRP gives the tear down of materials to be consumed and transformed for the production of stated finished products. In more dedicated production environments the production resource manager wants a dashboard that offers statistics, counts, and results of happenings in the production facility that sets targets and compares various periods to spot trends that need to be addressed and fulfilments that indicate the strengths the organisation is growing into.

The scheduler needs to understand the positions of the material planner with materials and production planner with methods as to how materials are drawn on the MRP and how production routings affect the timing of scheduled production. The scheduler needs to understand timing components of the production routing to understand how to adjust feeds to the MPS and the workability of the MPS in response to changes.

In certain cases where there is maintenance management function there is a maintenance scheduler role that schedules inspections and preventive maintenance activities to enable equipment availability when production is scheduled to them in their work centers.

Production Order Queuing

Every manufacturing production order is either triggered by a customer order or an internal requisition order. Production orders are initiated from incoming customer orders or internal replenishment orders. Customer orders come from the demand chain where customers demand product fulfilment from the enterprise. These orders place a demand on the enterprise production resources. When they arrive they are given estimated due dates at which they could be fulfilled. Such order fulfillment takes into consideration the enterprise capacity to fulfill the order, and these capacities are based on the total order within the period under consideration. The sales channel from which customer orders are obtained ensures that the enterprise is on the same page with the customer as far as specifications, technicalities, and customisations of the customer order is concerned to deliver good customer experience. Customer order quantities are the same quantities that appear on the production order generated from customer orders.

Ultimately, these orders initiated from customers are considered as make-to-order (MTO) production orders and they have particular characteristics in that the enterprise provides a level of customisation in retention of customer interest.

An MTO production order produces to fulfill specific customer requirements and as such has possibility to be defined differently for each of the enterprise customers. The interfacing elements of digital marketing, sales channel management, fulfilment management, and invoicing firms up the customer end of the enterprise in interfacing with the customer even up to the revenue cycle of things. However, when the order is absorbed it needs to be routed into production facilities digitally to engage production resources, which will grant access to the "five Ms" of production being machines, manpower, margin, materials, and methods. When the customer order migrates into a production order the material planner who attaches an appropriate BOM and the production planner who attaches the correct version of routing sheet will work on it. The BOM encumbers materials to be consumed, while the routing sequence provides activity timings to estimate the duration of production as well as machine, manpower, and method requirements.

MTO production orders queued to work centers encumber that work center for the estimated duration derived from the attached routing sheet when estimated as a product of estimated production time for one unit and quantity counts specified for the production order.

Replenishment orders also come from requisitions triggered by requirements for dependent materials due to MRP operations or EOQ settings in inventory control. Dependent materials such as spare parts, consumables, and components must be sustained to ensure smooth operations of the enterprise and their being held in inventory means a readiness to embark on production when production orders arrive work centers.

MRP being a material requirements plan is fed with BOM and MPS for independent materials to generate gross requirements. These requirements when phased out with on hand quantities generate order requirements for external entities whereas those that are meant to be internally satisfied are further progressed into make-to-stock (MTS) production orders. MTS orders being internally generated produces the sub assemblies and internally sourced materials which when completed are moved into the warehouse to handshake the revenue cycle protocols for its consumption fulfilment to customers.

EOQ settings in inventory control on the other hand keeps a watch at on-hand quantities so as to trigger a pre-determined reorder quantity whenever on-hand quantities reach the recommended reorder level. These settings are meant to secure a consumption regime that protects the safety stock settings from being violated so that the enterprise does not run out stock of critical materials though dependent. These are usually for consumables.

Provisions compiled into a production order are basically a BOM and routing sheet. These enable the appropriate work center to draw down materials and marshal out appropriate methods to engage the production process. The

BOM is the material mandate that the production order draws from the warehouse. The BOM is to be represented as quantity for making one unit of finished product multiplied by the number of finished products to be made, for each component. The BOM manifest is expected to have been exploded from MRP after an MPS had been run so as to assume good capacity plan and that requirements had been fully planned and smoothed to guarantee a stable MPS.

The routing sheet provides the machines, manpower, and methods to be applied to transform the materials drawn down into production. The routing sheet also provides estimates of time and costs involved. The routing steps are to be linked to BOM items so that those steps consume and transform the materials at those steps as required. The timing estimated at each routing step must be averaged down at execution with materials for record keeping for further understanding of the parameters of production when used to establish KPIs for an enterprise, cluster, or industry.

Production Order Execution

Production orders have attached provisions that aid their execution such as BOM and routing sheet. The first step in executing a production order is to mobilise materials from inventory with which to comply with the instructions embedded with the production order. Secondly, the attached routing sheet is interpreted as the production method required for completing the order. Executing the routing ensures that materials linked to routing steps are consumed in the process of routing execution. By executing the attached routing the ordered production is considered executed.

Work Center Job Management

Production orders queued to work centers enter a state transition from material receipt to warehouse release and back flushing. Queuing a production order to a work center engages work center resources so each work center must be equipped with appropriate resource capacity to deliver the production orders it receives. Materials in production enter a work-in-process stage when they enter a work center and must be further transformed into a final state of finished products. The transformation process engages machines, manpower, materials, and methods. Every work center that receives production orders must make capacity plans of machines, manpower, and materials. The mechanical capacities of the machines must be known to posses the expected duty cycle to support the yield expectations of the production order with respect to finished products and co-products that may result from the production order. The finished products will be released to warehouse while the co-products may be funnelled out as work-in-process for another production process where secondary processing may be required.

Multiple work orders may be queued to a work center in such a way that they are executed on a pre-established sequence such as work center priority rules as discussed in chapter 5 of this book under Job Management, such as FIFO, EDD, CJ, or SPT.

When a production order is queued the "five Ms" of production are encumbered in the proportion specified by the production order. Production resources are engaged and need to be gainfully managed to yield the best possible finished product counts. As less production orders are pending in a production facility more and more capacities are freed up for monetisation through inbound outsourcing strategies. If production capacities could be virtually showcased they could be sold virtually to enterprises that require those capacities and could rent those capacities for a fee.

Graphical Queuing of Production Orders

Production orders queued to work centers are best visualised in a graphic user interface (GUI) that schematically lays out work centers vertically to the left of the screen and time horizon horizontally to the top of the screen. Both work centers and time buckets should be scrollable to enable access to more planning resources on the screen. The current date by default should pan the time bucket to the left of the screen when the system is launched while the work centers could be represented by rectangular boxes and possibly alphanumerically and icon sequenced top to bottom. Planning production in this scenario could be as simple as dragging pending production orders and dropping them on work centers. Dropping a production order on a work center automatically secures a start date based on availability from the current date into a planning time bucket. This availability is based on any date that is free after the last production order end date or current date on that work center and should be drawn as a strip of rectangular schematic with initial bars depicting production order initialization details of order initialisation (OI) and order planning (OP). At the tail end of the production order strip are depictions of production order completion details of quality assurance (QA) and order release (OR). Such graphical workbench for production order queuing conveys a lot of meaning to schedulers and managers of the process in that it also gives a bottom status bar graphical total of capacity requirements so that the schedulers and managers are not left with too many moving parts. Ultimately, a graphical production resource management workbench will rapidly move production orders to work centers and often to work centers of the right capacity, as these issues can be graphically resolved more rapidly at queuing to resolve possible facility bottlenecks in future.

Production order execution commences with material allocation. Thereafter, machines are setup for material transformation, and then the appropriate manpower skill handles the material through the machines into the preferred transformation expected within the work center. While all these are going on the production order progress is graphically broadcasted as a sub-strip of the production order strip as a fraction of how much has been done compared to the total work to be done. When the progress moves to 100% the completion process is triggered for quality assurance and warehouse release. At this stage materials, labour, and overhead credit values represented as Work in process debit in the Finished Products WIP asset account can now be back flushed from the production journal as a debit to Finished Products inventory and credit to Finished Products WIP account, extinguishing the earlier WIP debit entered at initiation of production. Both Finished Products and WIP accounts are asset accounts with a characteristic that a debit posted to any of them increases their balance while a credit post decreases their balance.

Production Routing Management

Routings are the instruction sets written to draft task execution within production resources for identical implementation through multiple materials and components that make up the work-in-process (WIP) worked on. Routings determine 'What' to do, 'How much' of what, 'With which' machine, 'How' to do it, 'When' to do it, and 'Where' to do it to produce finished products acting on one material at a time until the finished product is realised. Routings, being an encapsulation of methods, sequentially structure the transformation of materials from one form to another by employing appropriate application of machines, manpower, and value adding operations. Routing implementations require good understanding of work centers, machines, and manpower operations as these are closely knit in getting the best out of routings for the required transformation of materials across work centers into the desired finished products. Routings are composed in routing sheets that are passed along with manufacturing orders to production facilities to engage the conversion cycle for the making of finished products. The better-composed routing sheets deliver greater production yield, efficient manpower deployment, most applicable machine engagement, and higher finished product throughput. Manpower must therefore be adequately skilled to interpret and operate production routings scheduled to their work centers for implementation.

Production routing moves materials from initial raw material through its transformation to the finished product, deciding the path and sequence of operations to be followed on the job from one machine to another, the manpower requirements, the time to be spent in every work center, and the estimated costs to be incurred through routing steps. Also factory layout, temporary storage of WIP inventories and material handling are major considerations in routing management.

Routing Composition

First having the end product in view and working the composition from a bottom up script combining first the base components creates routings, then including sub assemblies if any until the final product is realised. Good routing composition may also include inspection, testing, and quality assurance entries to assure that the finished products meet established standards and guidelines. Besides, material handling is a major component of every routing and must be adequately addressed within a routing composition to ensure that they are safely handled and do not diminish material value while in a work-in-process form before their eventual packaging for the marketplace. At creation, routings must address the needs of generic work centers such as carpentry, plumbing, electrical, mechanical, and chemical, as planned routings before being drafted to specialised work centers with specialised machines and manpower that must have combined basic work centers to derive esoteric work centers of live production facilities. A nicely composed routing sheet will have sequence numbers represent specific task lines that must be completed to deliver the objectives of the routing strategy. Planned routings could be subjected to further editing until the best is obtained by being a better communication of production framework to best deliver the finished product anticipated, at which stage it becomes a production routing which must have two things additionally. First, a production routing must be associated with a work center. Secondly, a production routing must be associated with a BOM that associates its embedded components to routing steps of routing sequences for consumption in production lines.

The fine points of creation of new routings are the estimation of cost and time for included tasks. A first estimate may not always be realistic but by averaging down from the cost and time layers of various tasks and material alternative options cut them out more and more accurate going into the future.

Digitally, routing records are composed on relational databases to be linked to BOM product structure and manufacturing orders electronically. In a non-digital world, routings are composed as routing sheets signed of by the production planner. Attributes of a digital routing record provides a data structure of a routing header in relationship with routing details as follows:

Routing Header	Routing Details
* Routing ID	* Routing ID
Routing Version	* Routing Step
Product ID	Task Description
Product Name	Skill Required
Product Spec	Machine Required
Planner	Work Center
Date	Duration
Remarks	Machine Rate
	Work Rate
	Setup Time
	Setup Cost
	Machine Run Time
	Machine Run Cost

The routing header record maintains one to many relationships with routing detail records using the Routing ID while the Routing Details maintains Routing ID and Routing Step as primary keys with the Routing ID being further used as a foreign key and will normally display and print as master-detail profile output.

Adding Routing Sequences

Routing sequences provide the task sets on which routings execute the production required to manufacture end products. Sequence entries as routing details transform materials from one state into the right formation required of expected finished products. When routing sequence entries are made they also define the degree of transformation required of materials, machine usage, necessary operation, skill set requirements, expected quality standards, inspection necessities, and the state the material must achieve at final routing step.

For the routing to effectively consume and transform materials the routing sequence must be linked to a BOM material item. A routing set must first be linked to a production BOM for its included items to be available to routing sequences. The sequence then can relate to one or more material items in the BOM. The relationship must specify the unit and measure of the material related to the routing sequence. The routing step must then specify the nature of transformation the related material is expected to experience based on which the material will be consumed at every engagement of the routing sequence. If the step uses machines the machine operation must also be specified in detail. This way the end state of the material after invocation of the routing step must be clearly specified with possibility of QA to ascertain the attainment of such material state before moving to the next routing sequence.

The task of each routing sequence step must specify duration in seconds, which must be tracked in performance improvement studies.

Linking BOM Items to Routing Sequences

The delicate balance of having BOM items linked to routing sequences is what makes routing sequences reinforce production operations into production patterns that precisely transform materials into desired finished products.

The specification for BOM item depletion in the routing establishes a consumption pattern for the BOM items linked to routing sequences. The degree of linkage specifies how soon materials are consumed on a live routing sheet and the rate at which materials are to arrive the work center from the warehouse.

Material consumption at exact manufacturing steps help derive the cost of the tasks involved and its importance in the value addition of the conversion process as well as the accuracy of the pre established value chain analysis to validate the reality of enterprise margin expectations by engaging the manufacturing process.

The linkage achieved must totally consume all items specified in the BOM otherwise the BOM will need to be specified to reflect the reality of its association with the routing to prevent inventory lapses such as shrinkages and shortages. It is such linkages of BOM to routings that validates the BOM for use as a key production resource to be utilised in processing MPS and MRP in production facilities.

Attaching Routings to Production Orders

For manufacturing orders to be executable they need to be perfected with routing sheets that convey the instruction sets required to streamline them through machine usage, manpower engagement, material transformation in production facilities. Routings profile the step-by-step task sequences required to fulfill the manufacturing operation outlined as a procedural framework for production. The task detail method serials in turn must have cascaded linkages for material consumption due to the BOM encapsulated in the attached routing sheet where some of the routing sequences are expected to draw down on material items from the linked BOM for consumption as manufacturing directions dictate.

The attached routing sheet of a manufacturing order will have entries for task details, machine usage, manpower intervention, and work center stopovers. Every task execution has a cost rollup that is expensed at execution and must be captured in manufacturing WIP accounts at closeout as credit postings to offset initial debit entries at commencement of production.

Work center crewmembers receive continuous training and retraining on how to pick and unpack materials specified in the BOM to effectively operate the routings and where materials require special handling, such as toxic or corrosive materials, they are directed on the proper utilisation of personal protective equipment (PPE) for their safe handling.

Routings attached to manufacturing order exposes the BOM linked to the routing sheet and its material items linked to the routing sequences of the routing sheet such that an inheriting manufacturing order is readily executable based on the percentage of possible material consumption evidenced in the attached routings.

Routings attached to manufacturing orders determine how materials enter the work centers from warehouses, and it establishes the replenishment patterns required to maintain optimal production.

Work Center Management

Manufacturing production takes place in work zones of industrial areas specifically designated as work centers where machines are installed and manpower assigned to execute manufacturing orders leading to production through transformation of material components into finished products. A work center will consume materials in a conversion cycle where these materials are transformed into finished products, it holds machines that are arrayed and deployed for transforming materials in the conversion cycle, houses manpower that are skilled in the interpretation of production routings and material handling for adequate method management, quality control, inspections, testing, and quality assurance, and being a production unit they are able to receive manufacturing orders that they interpret into manageable tasks and operations to be sequentially executed within their capacity and resource constraints.

Work centers are expected to be conformable, productive, and sustainable through superior facilities management that contributes to the organization's bottom line. Work centers must be top on safety, a safe work center is good business, and safety programs are to prevent work place injuries, illnesses, and deaths, as well as the suffering that goes with poor workmanship. Making work centers safe is to secure the comfort and morale of manpower that drives them to produce the finished goods the enterprise need. Good work centers must be efficient. Efficiency makes the work scheduled to them more manageable and approachable.

Work centers consist of machines and manpower on which routing operations are dispensed. Machines transform materials for value addition in a conversion cycle with recognizable margins using a machine cluster or an assembly line. While manpower operate the machines and handle materials that enter into the conversion cycle for value retention while in a work-in-process form so that there is no diminishing in value, specialised manpower are organised as crewmembers or shift labour, depending on what is most appropriate for the work center. Crewmembers spread out and tidy the work sequencing that places order on how tasks are completed, while shift labour take turns round the clock in a shift like two or three shifts daily (24 hours) to provide manpower for continuous manufacturing processes that must be manned.

Work centers derive the production timelines that production orders boast of as these timelines are based on the resources domiciled in the work centers.

Work Center Grouping

Mapping work centers into groups gives them greater capacity to take on bigger jobs with a unification of manpower and machines. Grouped work centers elevate resource capacities making such clusters improve throughput by the improved capacities they achieve. Work centers with identical characteristics are often grouped for effectiveness, simplifying monitoring and effectiveness in the use of their machines and manpower, placing them in a group streamlines their operations as a bigger production field. Grouped work centers will usually be replenished from a common inventory pool for effectiveness. Routing operations are to be directed at work centers and to dedicated machines in particular. When routings are directed at work centers they encumber the machines and manpower crewmembers of the work centers. Where work centers are operated on rental machines and contract labour, scheduling work to them implies that those incidental costs must be born to operate the work centers and absorb the costs into WIP accounts. In work center groups the total capacity of the work center group is the sum of the sum of the individual work center capacities.

Efficient running of work centers is in the management of manpower and machines in readiness to receive manufacturing orders into the work centers. Their status is based on how ready the manpower crew is and the state of machines in the work center. An arriving manufacturing order will provide manpower with engaging instruction sets and routing sequences to be followed in manufacturing operations as well as the reporting required to maintain the document travel across work centers. For materials there will be some authorisations required to retrieve materials from warehouse if they are not pre-released to work centers prior to manufacturing order arrival.

The machines defined in routings are expected to be those of work centers to complete the sequence steps specified in the production routings as described, this means that machine and manpower are encumbered for back flushing through the engagement of the work center and manufacturing order.

Digital Work Center Records

Data structures for work centers store information on location, definition, capacity, machine, and costing, and timing. Work center records are used for scheduling, costing, and capacity planning. For scheduling, accurate and correct work center records provide data that helps estimate the timelines of work scheduled to them using the specified working time of the work center which could be 24 hour round the clock or shift work time in 3 shifts of 8 hours each, dual shift of 12 hours, or single shift of 9 hours daily based on the total estimated from the production routing.

Work center machine setup costs, running costs, and labour rates determine work center operation costs. Product of work center machine hours and the number of machines as well as manpower hours and number of manpower determines available work center capacity. When ponding manufacturing orders are scheduled to work centers the constants stored in the work center records automatically calculates the parameters based on production variables in the manufacturing orders so scheduled. Resource planning with work center systems precisely calculates the manpower required to meet daily, weekly, or monthly manpower schedules. Resource planning smartly manages the manpower based on qualification and the relevant skill sets.

In processing manufacturing orders work centers receive work assignment, not just that but processing production routings also require that routing steps be assigned to work centers from a drop down list box, and further to that a list of available machines in the work center will be further refreshed in an adjacent drop down list box for assignment as well. This makes the work center record critical to manufacturing operations.

Dedicated Manpower

Work center operations are carried out by specialised manpower. Depending on the required engagement of the work center there is a pool of standby manpower that carry out basic assignments such as cleaning, communication, storage, material picking, and material putting away. Then there is the skilled manpower that must operate the relevant production machines. By working as a team the crew share knowledge and routinely carry everybody along so that their weak points are not allowed to get in the way of being a production bottleneck. In larger work centers desirous of work center safety a special squad must be responsible for sanitation, safety, and health of the crew. Modern work centers even have provisions to sustain the physical vigour of special crewmembers for specific tasks.

Work Center KPIs

Work centers being critical in manpower deployment for enhanced productivity, material handling for operational efficiency, and assignment of routings where enterprise production methods and strategies are realized need to be monitored to ensure they deliver enterprise expectations such that business metrics used by executives and other managers to track and analyze factors deemed crucial to the success of the organization's production strategies are sustained. Effective work center KPIs focus on the business processes and mechanisms that management sees as most crucial for measuring progress toward meeting strategic goals and performance targets of work centers. When work center dashboards project the collective KPIs of an enterprise' production the organisation knows what to do to sustain them and pool the right talents to the enterprise to support the needs of their work centers while holding onto the right squad to execute manufacturing orders such that gallery of knowledge, skills, and abilities give the right mix of manufacturing acumen to the competitive advantage of the enterprise.

Work center KPIs are about measuring the efficiency and effectiveness of material handling, manufacturing assignment, and manpower. Material handling effectiveness is measured by measuring the efficiency of handling a unit weight through a unit distance, layout efficiency, these are improved by prioritising simplification, ergonomics, utility, and systematization of materials. Manufacturing assignment effectiveness measures routing flexibility, facility efficiency, . Manpower effectiveness is measured by using rating scales, establishment of team performance objectives, and by quality and quantity of produced units.

Work Center Accident And Injury Prevention

No matter how busy crewmembers are the workplace must be kept clean. Water, oil, tiny metals, grease, ingredients, and even sand can create a substance harmful to manpower motion in the workplace, so work centers must be kept regularly clean, possibly with a sanitation squad. Machines and tools must be kept out of the way when not in use so there's no danger of other crewmembers tripping over a loose machine or tool. With those slippery surfaces of industrial work centers anti slip mats should be made profusely available and ensure that floors are mopped regularly and free of slippery substances such as standing water. Moreover, slip resistant safety shoes must be generously provided as a workplace protective equipment to protect against the hazards of limb damage in work centers that are prone to slippery.

Generous use of personal protective equipment (PPEs) must be enforced to secure the health and safety of manpower limbs from damage to avoid disputes and preventable liability settlements. Teamwork must be encouraged in events of lifting heavy objects to prevent individual musculoskeletal injuries, which will cause loss of time and money.

Climatic controls must be put in place in most circumstances to secure the comfort of manpower from heat or cold generated from the work environment, and that complimented with appropriate hygienic fabrics and ergonomics customised for the work center. In places where noise is an issue, earplugs must be provided as an optional feature, alternatively special earplugs can be provided that limit noise prevention to harmful noise but not those of crewmembers so that instructions can still be heard audibly but harmful decibel noise are prevented from passing through.

Work center workflow is a very important consideration and must be critically considered in the placement of machines and tools. There must be an efficient layout that lets work move easily from one machine to the other and that tools can be efficiently picked, and effectively handled, on an as need basis without disrupting the flow of work. Work center layout must take into consideration how everyone moves and works in them to build a proper layout of machine, tools, order of routing sequences, and manpower that makes everything work efficiently and safe. It should be thought of ahead of time what hazards pose a risk to manpower and let it be figured out what equipment, accessories, consumables, and PPEs need to be provided to prevent injury.

Work Center Communications

Work center communication is critical to get the job done. With verbal, non-verbal, and visual communication work centers are able to release critical information concerning expectations from the workforce as well as the progress and routing of work in process materials. When work centers engage high impact communication responsibilities are assumed to ensure that work centers live true to their purpose and increase productivity. Teams need communication to get the job done, particularly when passing work-in-process to the next team in line in the next work center. Such passage may include signing off on their portion of the routing sheet, recording man-hours, machine hours, and possible shrinkage. This form of detailed communication is a commitment on the part of work center leadership and takes responsibility to admit; yet it is such indices that enter the production manager's dashboard as a show back of how the enterprise is doing. Communication must be effective within every organisation to keep all eyes on the ball as communication gaps must be avoided by all means because it leads to confusion, production losses, and lost time. Effective communication strengthens manpower cooperation and commitment to enterprise goals, ensuring that deadlines are not missed, and that productivity is elevated as agreed. Crewmembers and their supervisors interact clearly and effectively through verbal and non-verbal communication. In verbal and non-verbal communication care must be taken to understand the background of the recipient of the communication as not doing this may lead to communication barriers because the message recipient may see differently from what is communicated and such communication barriers may prove difficult to resolve.

Simple, clear, and precise communication is required to ensure everyone reasonably absorbs communication in work centers. Visual communication aids understanding in the presentation of vital information in the workplace and help to create a tension free workplace.

Today's workplace communication involves electronic means of communication such as internet, email, and social media. It is important to keep in mind that sending an email, a letter, or a social media post does not necessarily mean that communication has taken place. Only when a message has been sent, received and understood by the intended receiver, it can be said that communication has occurred.

Work Management

A well-configured work center coordinates production orders for transformation of materials into finished products and maintenance orders for machine and equipment reliability. Work management for production of finished products is by scheduling of production orders to the work center where the order is allocated resources and in its teardown the attached BOM and routing are used to conduct the transformative work of value addition which when completed with QA the finished products are moved to warehouse storage.

Work management for maintenance orders, on the other hand, when scheduled to the work center is to be approved and assigned by the work center's supervisor to a technician who performs the work while the supervisor monitors the progress and closes the maintenance order when the desirable status of the facility has been reached and further inspections conducted to asses and survey the state of the facility reached. As opposed to production orders, maintenance orders achieve value preservation rather than value addition.

Material Management

Having a good material management strategy help enterprises hold capacities to meet production schedules, customer service, merchandise fulfilment, or even running the enterprise inventory in various structured storage. Materials can be accounted for both in traditional storage such as warehouses, retail outlets, and manufacturing plants, or non-traditional storage such as in-transit or in-waiting storage. Materials are critical production resources, and as such they present at various times for transformation, usage options, and value addition as a spectrum of materials such as: consumables, components, finished products, ingredients, kits, merchandise, raw materials, spare parts, sub assemblies, and work-in-progress.

Consumable materials are held in warehouse to be used up in providing services or for the performance of operations. Components are held for production of end items such as finished products, kits, merchandise, sub assemblies, work-in-progress, or spare parts. Finished products are a final combination of raw materials, components, kits, sub assemblies, or WIP in a form that is ready for end users as end items produced for storage (MTS) or to-order (MTO). Ingredients are a form of raw materials held in inventory in a catering facility for the production of food as end items. Kits are a combination of items held for production of spare parts, components, sub assemblies, merchandise, WIP, or finished products. Merchandise is Item held for retail operations. Raw materials are held in inventory for consumption in the production of finished products. Spare parts are items held to be issued for maintenance operations. Sub assemblies are Semi finished output of WIP operations. While work-in-progress (WIP) items are intermediate process output of manufacturing production as materials that have encumbered the manufacturing process and are therefore pending for subsequent production activities to be carried out on them.

An enterprise with greater grasp of the best of material management within its capacities will be more appreciated on the supply chain as there is bound to be considerable penetration of the value chain of any industry with quality materials.

Material management is used by enterprises to plan, organise, and control material consumption. In today's production environment the management of materials across its broad spectrum is done by considering how materials will be consumed by production resources through the mechanism of the material requirements plan (MRP) for a more data-centric approach. Material organisation is through the laying out of storage strategies for material accumulation and retrieval in a warehouse structure. Material control is by regulating, parameterising, and monitoring material movement by use of an inventory control system.

Enterprise management of materials involve material warehousing, material planning, storage management, and inventory management. These help the enterprise to keep its eyes on critical materials required to make its MPS come alive and its MRP to persistently bring in the right requirements to engage production resources.

Material Warehousing

A warehouse is a game changer for the enterprise as a critical asset for material value retention. A warehouse is a structured storage solution and environment where materials are accumulated in storage systems for value retention and enhancement to support procurement, fulfilment, and value addition activities from the supply chain, for the sales channel, and to support an efficient production system and all enterprise sites for operational excellence through efficient storage and retrieval processes.

Warehouse management is the administration of adequate clearinghouse to support all enterprise operations with efficient material storage and retrieval strategies. Warehouse management is expected to provide handling capacity to enterprise material consumption signatures for a wide variety of product families for fulfillment and replenishment operations with adequate space for material storage footprints. Consumption patterns vary depending on prevailing fulfilment, replenishment, and production patterns. With MRP or economic order, or whatever pattern there is a stream of defined inflow of inbound materials. In the same vein, other ERP modules based on the operations of the enterprise define the outflow pattern of inventory. The management of warehouses marshal detailed strategies for enterprise inbound and outbound logistics by deliberate warehouse layout and racking structures to facilitate material value retention in holding areas. Warehouse management sets up the objectives, layouts, shelving, and sections that become structured holding areas for materials.

Accumulating materials into a warehouse is expected to retain its value so it must be done within good material handling guidelines. Its retrieval must be seamless without compromising the warehouse integrity so it must be done not just with efficient retrieval but also safely with materials securely held in primary, secondary, and even tertiary packaging that are compatible with material handling devices.

Enterprises must ensure that warehousing policy is built around its material handling procedures for value retention, as not doing so is to subject materials to uncontrollable value degradation and unsustainable material handling approach. Warehousing does this by offering mechanisms with which to accumulate and retrieve materials in storage systems and retrieve them at will. The material handling protocol and the housing provided by the warehouse retain the value of materials placed in the warehouse for the duration it is designed. There are types of warehouses for different kinds of materials. For common merchandise materials it could be just about any building depending on the functionality required, but for fluids as used in the petrochemical industry it is must be a tank farm or plastic drums. For seeds harvested from crop farms storage silos are preferred. For frozen food materials such as chicken and beef, electrically sustained cold storage is the preferred option. From materials just being kept in a shed enterprises realised that materials are a major asset that need value retention because degradation could be an economically harsh reality and so the successful enterprises built their profitability around the best warehouses they could afford to the extent that they have built smart warehouse being an automated warehouse offering a way for faster turnaround in accumulation and retrieval of materials in lock-step with their operational excellence.

To the flourishing enterprise warehousing offers very good appeal, for instance a warehouse is an option as a fulfilment center for merchandising companies that move materials across wide geographic locations to better manage their outbound logistics or distribution logistics to reach their customers for better customer service. Also for government and large companies involved in corporate social responsibilities (CSR) a forward logistics base warehouse helps with service delivery to distribute materials or get work done, sometimes with involvement of inbound and outbound for forward logistics, and material recycling structures for reverse logistics.

Warehousing secures materials from degradation. Materials placed in a warehouse are secured from the harmful effects of weather. So long as a mitigation plan is in place and warehouse manpower is enlightened with the right tools to work an emergency response plan must equally be put in place to ensure asset security. The warehouse must be weatherproof and arranged for maximum productivity. Warehouses need to be secured places so that materials in them can be secured too. The key here is to have appropriate material handling guidelines in place so that materials can be easy to store, find, move, and ship out of warehouse with minimal or no damage.

The warehouse should be secured with cascaded security measures such as: entryway security doors, access control systems, closed circuit TV (CCTV) cameras, good lighting system, security patrols, and alarm systems.

The six activities involved in a material warehouse are receiving, put-away, storage, picking, packing, and shipping. When materials are received they are placed in an in-bound staging area while supporting documents are crosschecked before accepting liability on behalf of the enterprise, else they are considered a receiving exception that must be escalated to top management for resolution. Put-away follows the receiving of materials in assigning them their proper shelving through proper documentation. Storage is the physical shelving of materials through appropriate material handling. Picking is the assembling from the shelves to the out-bound staging area of materials required to fulfil internal or external requisitions. Packing is the boxing of picked materials for delivery based on delivery constraints. Shipping is the delivery of ordered materials to external entities following a predetermined route pattern by the enterprise.

Material Planning

An enterprise plan for materials involve the requisition blueprint made for dependent materials required for production of finished products outlined in the master production schedule (MPS) which is an embodiment of customer demand passed for make to order (MTO) production and material requirements for make to stock (MTS) production. The specific mechanism for driving the material requirements plan (MRP) is a step ahead of inventory positions and the MPS in that it takes vendor lead times and material safety, which is the quantity that must be in stock when an order is placed into consideration to ensure the enterprise never runs out of stock of critical materials by using an economic order quantity which is basically a saw-tooth formula of material consumption in the enterprise with a consumption gradient that swings between turning points of maximum holding parameter and minimum holding levels while sloping along a horizontal duration axis, where the reorder quantity must return the stock to its maximum level upon vendor delivery that was triggered by the reorder level. Critically, the vendor's lead-time must be contained within the specified safety period defined for the stock item for the formula to work. The MRP when triggering material requisitions for dependent materials will set the order quantity based on the reorder quantity specified for the material and will attempt to justify its usage on the MPS but will likely enter into an iteration to establish multiple reorder quantities for the MRP output if the MPS enters into a race hazard due to insufficient quantities, this it must do to protect the integrity of the MPS as an overall master scheduling tool for the enterprise.

Planning for materials must further be with adequate documentation. Efficient documentation of material attributes and characteristics stimulates material optimization and consistent usage, storage expectations, and innovations around materials with clear-cut data handling precision. Maintaining a consistent flow of materials for production is the objective of a good material planning.

Storage Management

Material storage management is structured aggregation and codification of storage systems for efficient accumulation and retrieval of materials from storage.

Appropriate accumulation of materials in storage are required to retain the values of these materials by limiting people access to them while retaining them in easily identifiable storage locations and systems. Material storage is made in piles at whatever location to make way for more materials, this concentration of materials in storage help to limit the holding costs associated with storage. To help achieve material concentration bin locations need to be labeled for easy access, both in numeric and alphabetic combinations for legible traceability.

The storage management system must protect materials it holds from value deterioration. Rather, the value of the materials accumulated in a storage system must remain the same at retrieval as it was at accumulation. Value retention is a critical justification to hold materials in storage systems.

The storage system adopted must offer protection to materials kept in them, both when stored in piles and when loosely stored. The storage must be organised and neatly arranged for easy retrieval and to allow room to navigate with material handling equipment in use in the particular site. The racks and shelves must be rigid and firmly secured to the ground to prevent them from caving in or falling over when holding materials.

To improve storage systems, weight of materials must always be taken into consideration while accumulating materials into storage. Care must be taken to use specialised, accessible, and well-labeled storage racks to improve accumulation and retrieval efficiency of the warehouse. Interlocking racks could further improve rigidity of the storage system as rack weights are transmitted evenly to the warehouse ground rather than hanging weights. With a good storage system based on efficient facility layout the flow of materials is responsive to the needs of the enterprise and materials can be accumulated into the storage system and retrieved at will to serve the activities of the organisation with good material handling.

Storage management done right reduces manufacturing costs, secures future value of assets, lowers shipping and receiving exceptions, shortens delivery time, fortifies manpower responsibilities, develops vendor loyalty, improves customer satisfaction, and adopts automation that reduce or even eliminates cumbersome manual recordkeeping and reporting

Inventory Control

Inventory control is chronological documentation of material movement and valuation details for up to date information on material status.

Inventory control documents the movement of materials within all enterprise sites with a view to having an up to date running balance of materials. The balances provided by inventory data triggers sourcing and procurement activities for material accumulation into storage systems. All types of inventory are important and impact significantly on enterprise bottom line. The inventory policy could be based on EOQ, Kanban, JIT, or MRP, for efficient management and for demand satisfaction. The optimum inventory policy should provide greater levels of customer satisfaction and minimization of the cost of investment tied up in inventory with FIFO, LIFO, weighted average, and specific id valuation models to give much needed accounting to materials. A **3-Bin Kanban Inventory Strategy** better manages material logistics by integrating the 3 key components of the requisition variable being the shop floor, the enterprise warehouse and the material's preferred vendor. By sharing common inventory control parameters between these 3 each is able to peek into the consumption pattern of the entity it manages and respond accordingly. The 3-Bin Kanban empowers virtual requisition (VR) to ensure the enterprise never runs out of critical materials as the warehouse and preferred vendors has them covered electronically.

Standardisation of Material Management

The international Standards Organisation (ISO) being a global regulator in all fields of development publishes a number of standards to benefit organisations and advance global development orientation to keep organisations on the same page of best of breed judgements in whatever area of development has released a number of guidelines in the area of material development that can help anyone involved in material management function – whatever their background and size – establish, maintain and continually improve effective and functional material management and governance processes.

The ISO standards applicable to material management are listed below:

| STANDARD | DESCRIPTION |

ISO 9001	For quality management systems. Provides guidelines for maintaining high quality materials in storage
ISO 13485	Governs review, approval, and continuous monitoring to secure conformance to medical industry safety and quality regulatory requirements
ISO 14001	Provides a framework that an organization can follow for environmental management best practice towards minimization of environmental footprint and diminishing the risk of pollution incidents in the handling of materials, promotes resource efficiency, waste reduction, and recycling in warehouse operations.
ISO 17025	Standardizes testing and calibration laboratories in the warehouse for testing materials. The standard provides a set of requirements the laboratories are to follow to show that they operate a quality management system and that they're technically competent to do the work that they do
ISO 45001	Governs occupational health and safety standards for the warehouse to reduce workplace risks and create better, safer working conditions
ISO 22301	Protects societal security standards with guidelines on what to do when things go

	wrong
ISO 27001	Provides specifications for information security management system (ISMS) to protect information stored in the warehouse database with a framework of policies and procedures that includes all legal, physical and technical controls involved in information risk management processes

Maintenance Management

Assets are the lifeblood of an enterprise; an organisation is in operation based on the soundness of its assets. An organisation's assets help it generate revenue cycles upon which its capital is built. However, in the normal course of business life assets are known to depreciate in value, physically deteriorate, and tend to become liabilities economically if maintenance activities are not adequately engaged to reverse the tide of deterioration. If asset depreciation takes a toll on the enterprise the assets becomes a liability and tend to pile up expenditure cycle transactions that chalks up debit expenses against the organisation due to malfunction, while also limiting organisational capacity to engage flourishing conversion cycle activities that have the possibilities for heading the organisation in the direction of feasible revenue streams, these debit accumulation due to equipment malfunctions erode the organisation's capital and a reason why asset management or maintenance management is strategic for the growing enterprise. Todays enterprise is tilted towards asset intensive manufacturing and as such the life cycle of its physical assets must be favourably maintained to provide the enterprise with the best service life possible if the enterprise must remain afloat long enough and in the process reduce total cost of ownership (TCO) over the asset lifecycle in favour of flourishing revenue cycles.

Assets are economic tools that provide conversion cycle availability that is influenced by various economic factors. The strategy for effective maintenance is to limit downtime in favour of maintenance costs. Maintenance costs are not cheap; they involve on-boarding manpower of the right skillset, stocking up on spare parts, implementing safety plans, tool acquisition, and commitment to reporting, notwithstanding the requirement to have a maintenance policy in place.

Manufacturing production engages machines, a lot of them. Making production work wears out machines, and they need to be renewed every time they get worn out so production can get back as scheduled. Every machine wears out as they are used; some even wear out just by being there.

Maintenance helps an organisation absorb operational risks by ensuring that machines are up and running and able to deliver production orders scheduled to them and equally taking out time to deliver the right talent to operate these machines appropriately so their need for maintenance are reduced to the barest minimum.

Maintenance management is a methodical and systematic approach to planning, organizing, monitoring and evaluating activities and their costs that improve the reliability of assets such as buildings, machines, and equipment to prolong their individual service life.

Enterprise assets encompass vehicle fleet, electrical systems, mechanical systems, buildings, cooling systems, piping systems, various machines, plants, and equipment used in production. They are all bundled in enterprises as assets. Maintenance of these assets involves functionality checks, servicing, repairing, replacing, or upgrading of necessary devices, and all these can be supported with corrective, preventive, risk-based, and condition-based maintenance to improve their service life and limit downtime to the barest minimum. Maintenance management is undisputedly considered an industry best practice that leads to superior performance.

Maintenance Planning

The first task in maintenance is setting out what needs to be maintained and what procedures to be adopted as well as by whom. Thus. Maintenance planning brings assets into the fold of a proper maintenance program, bringing them into routine inspections to observe their performance and kick in work activities to bring them to a serviceable life when they deviate from expectations.

Maintenance planning delivers procedures to be followed for each asset family, providing an outline on the purpose of maintenance activity to be done. Reference document to be relied on, then the tools and spare parts required. Conditions to be maintained, such as safety and operating conditions must be stated. The required location for maintenance activity must also be stated. All these are to be dispatched with maintenance work order whenever maintenance of each asset family so stated is ordered as necessary information for the maintenance crew.

The logic of planning is that inspection procedures are dispatched for all asset families and based on inspections observations are recorded that describe asset conditions and recommends remedial action to be taken to prevent further deterioration. Inspection findings will often specify the priority of remedial action and based on this the planning function can schedule preventive maintenance to prevent further deterioration of the asset.

Much of planning is to establish a baseline condition for all machines in the facility, such that any deterioration below such set standard must be immediately remedied. All machines in the asset portfolio must be adequately tagged and inspection frequencies must be established at which routine inspections must be carried out. Beside routine inspections, factory manpower should be empowered to report asset condition necessitating inspection based on criticality to production. The objective of inspectors is to observe the asset condition and report such conditions to planning. Inspections then move in detail to routine maintenance, pre start inspection, operational checks, and troubleshooting in an attempt to assure asset reliability before documenting maintenance requirements. The inspections then determine follow up activities to be undertaken, whether preventive, predictive, or corrective maintenance. Planning could also set due date for future inspections based on predetermined settings for asset families and what to be observed from subsequent visits. The overriding objective of inspections should always be to deliver greater asset reliability.

Preventive Maintenance

To assure production facilities of availability of value adding assets a periodic routine is set aside for the enterprise asset portfolio so that assets are up and running and that total cost of ownership does not extend beyond what is originally planned by setting aside certain periods of activity to ensure that breakdowns and malfunctions are prevented from occurring so that equipment availability means they can receive production orders and that the conversion cycle enables succession into a robust revenue cycle.

Time-based preventive maintenance requires establishment of fixed interval frequency to apply documented maintenance procedures with the right tools, the right consumables, and the right craftsmen. Such discipline is what planning does in inspecting and observing asset conditions heading towards maintenance routines to ensure quality reporting when it's all done.

Usage-based preventive maintenance observes asset activity and triggers preventive maintenance based on realized exertion based on pre-determined predictive parameters. Predictive parameters are often based on benchmarked meter readings to trigger preventive maintenance. Usage based preventive maintenance avoids over maintenance of lesser-used assets, saving financial resources.

Condition based preventive maintenance monitors certain indicators showing signs of decreasing performance or a forecast of future failure to determine maintenance activity that need to be done to return the asset to required performance. Condition based maintenance relies on acquisition of data and information as well as transmission and storage requirements to extract relevant information to understand the actual condition of the asset and predict its useful life. The objective of condition-based maintenance is to keep the asset within safe limits away from lower and upper limits to secure the value addition capacity of the asset.

Predictive Maintenance

Predictive maintenance is monitoring the condition of assets in an attempt to forecast the onset of problems to prevent devastating failures. Regular monitoring of asset condition and operating efficiency supports data gathering that establishes a reduction in the possibility of asset breakdowns. With the optimisation gained in total asset operation the cost of maintenance is significantly reduced, improving product quality, productivity, and profitability. Predictive maintenance is aimed at reducing costly, unexpected breakdowns through real-time asset condition monitoring.

Why Predictive Maintenance?
Predictive maintenance enforces compliance with safety directives, pre-emptive of corrective actions, and prolongs asset life. By gathering data on asset performance the environment is set to get the best from the asset, thereby complying with laid down safety directives of the facility. The up to date information provided about the asset performance negates the need to subject the asset to corrective actions, as this is not necessary since the asset is considered to enjoy optimal performance under a predictive regime and in an atmosphere of data that is ranked as signifying asset dependability for production operations.

Forecasting of Predictive Maintenance. The main forecasting basis in predictive maintenance is based on early fault detection strategies to predict the most likely time for breakdown and failures. With real time monitoring of asset condition a lot is acquired in terms of information content capable of showing the trend in asset condition that could trigger schedules for preventive or corrective actions based on predetermined benchmarks. Most importantly, maintenance forecasting provides credible asset intelligence for estimating overall downtime and proactively budget for maintenance in advance.

Undertaking Predictive Maintenance. The workflow of predictive maintenance follows a rather straitjacketed approach first by determining in advance what asset failure modes to track in its monitoring process. Then a monitoring frequency is established for its data gathering, after which the environment is established for asset condition monitoring with real-time data gathering. Reports of the monitoring are kept accessible to relevant teams. Whenever the data indicates an adverse condition it is planned into a maintenance work order with enough entries for planned work date, consumables, and spare parts required. Otherwise, asset conditions should continue to be monitored until adverse conditions are detected.

Predictive Maintenance Tools abound to make the activity manageable. All predictive maintenance tools provide real time condition monitoring: whether for vibration where vibration analysis is provided, for ultrasonic where ultrasonic analysis is provided, for infrared where infrared analysis is provided, oil where oil analysis is used, then there are motor circuit analysis, and laser shaft analysis. All these tools provide real time data on asset condition, are able to build a trend, compare acquired data to baseline data and decide what kind and level of trigger with which to indicate necessity for maintenance.

Data Analysis is what is done with acquired data to decipher the information content of big data on asset condition in arriving at specific triggers. Autonomous operation of data gathering is conducted in real time as a strategy that applies supervisory control and data acquisition (SCADA) on an industrial data pathway with complementary accessories and transducers efficiently laid out for long term planning. Running time series and decision trees on these data help to fish critical data points about asset condition with which maintenance planning is alerted. For image applications, the image grabbing, image processing, and image analysis aspects in weighing in on big data must be able to deliver distinct results for decision support on the sate of assets under observation.

Scheduled Maintenance

Scheduled maintenance is activity that is pre-arranged and assigned to a maintenance crew for resolution at regular intervals of recurrence through inspections, amendments, regular service, and planned shutdowns. Scheduled maintenance encompass routine maintenance done to improve asset reliability, pre-start inspections carried out to observe assets after shutdown events, operational checks to observe the performance characteristics of assets, and troubleshooting aimed at seeking out fault causes. Scheduled maintenance harnesses digital record keeping tools to create and monitor asset schedules at which they are subjected to maintenance routine to assure their performance in the setting where they are deployed. Performing scheduled maintenance prevents long-term deterioration and costly damage, ensuring superior asset reliability and elimination of costly downtime.

Pre start inspection is embarked upon at all scheduled maintenance to confirm the asset status from where it is, to determine in advance what the maintenance expectations are, and to confirm that the work can be carried out in a safe environment that will not result in lost time injuries.

Operational checks help to reveal the characteristics of the asset in ideal operation and how maintenance activity can restore this state to the benefit of production.

Routine maintenance is the actual maintenance performed on the asset and requires consumables, spare parts, and the right tools to have been encumbered for this operation.

Troubleshooting is the post maintenance operation that is carried out to confirm that the maintenance operation was indeed carried out by actually putting the asset to operation and observing its performance characteristics based on specific operation to which it is engaged to task its mechanisms for benchmarked functioning. The outcome of troubleshooting is what determines if the asset was adequately restored to its operational characteristics and if there are notes for future inspection and the next maintenance routine.

Corrective Maintenance

Assets often enter into a state of unexpected unforeseen breakdown due to a number of factors, such as negligence, bad working conditions, and overload. When assets in use break down they must be immediately brought back into usefulness by engaging corrective maintenance routines to rectify the problem that caused the breakdown and return it to its required condition in readiness to take on the workload that is usually scheduled to it. Corrective maintenance kicks in when equipment failures occur to correct the anomaly and return the equipment to productive functioning. Problems are what kicks corrective maintenance into action to certify optimal functioning and rectify a nagging malfunction. Before embarking on corrective maintenance there is need to isolate the equipment to ensure a safe working environment, and then the fault needs to be isolated to be focused on with the best of corrective approaches.

The speed with which a broken down machine can be returned into service notwithstanding, what is more important is the cost of maintenance as corrective maintenance tend to plunge finances into cost overrun and could be avoided for exorbitant cost even if it becomes a bottleneck to production and could lead to a piling up of broken down machines that will inevitably require a more expensive turnaround maintenance (TAM) of asset groups to bring a facility back into operation. Corrective maintenance is a proactive maintenance management; its goals are to eliminate breakdowns and deviations from optimal operating conditions so that critical production equipment is optimised for greater throughput. The in-service approach of corrective maintenance could be subjected to immediacy or deferred activity, depending on the overriding enterprise policy.

Following a failure, corrective maintenance observes; isolates the failed part, orders a replacement through the procurement system, then the faulty part is replaced, after which functions are tested, and finally the equipment is rolled back into operation. The objective of corrective maintenance is the mainstreaming of reliable assets in the facility on a case-by-case basis; it is maintenance for reliability, a most enlightening approach to maintenance with greater insight into the nature of machines and patterns of reliability that helps mitigate the negative effects of breakdowns and the application of best of breed planning tools to get machines up and running in a straight-forward, straight-up, and straight-through maintenance strategy that releases production resources to operate at full capacity.

Maintenance Materials Management

Planning for maintenance require detailing all the materials required for performing the maintenance work, a kind of MPS-MRP approach to maintenance practice that explodes material requirements for each type of planned maintenance activity. The procurement mechanism must kick in to ensure that required materials are held in inventory prior to scheduled work by engaging the appropriate order policy with preferred vendors. That way when the work is scheduled there will be materials such as spare parts, consumables, and kits that could be encumbered to deliver scheduled maintenance.

To do this maintenance needs critical material information such as real time material information, equipment data, inventory balances, and projected material delivery dates. Digital online material information required by maintenance are for planning activities, in making material choices for maintenance activities, planners need to know the on-hand situation of materials. The necessary material information required by maintenance planners; part numbers, part description, on hand quantity, quantity on order, and alternative part numbers, help to put materials in readiness for maintenance, eliminating lost times and material wait times, and also that materials are not haphazardly issued but to requisition orders that state the requisite machines that will expend them, that control with requisite machines is a major innovation in data management that can significantly limit maintenance inventory costs due to material shortages from diversion or misappropriation.

Equipment or asset information required by maintenance must be adequately represented in the asset inventory. An organisation's asset portfolio must be efficiently documented to effectively support implementation of asset management plans for performance evaluation and improvements through prioritised investments to benefit the asset lifecycle, knowing when to acquire, enhance, or dispose non-serviceable physical assets.

Maintenance Control Activities

The objective of maintenance is to deliver reliability. Planning for maintenance must align with long-term organisation objectives and be positioned for risk mitigation in establishing a resilient, reliable, and sustainable asset portfolio.

By analysing breakdown causes maintenance manpower are able to be less fully occupied with maintenance to devote more time to research solution strategies for greater asset up times and less of breakdowns and overtimes, with an independent research activity manpower could be dedicated to finding out the real causes of breakdowns and submit how best to go about corrective, preventive, or predictive maintenance in delivering proactive solutions to prevent breakdowns and malfunctions.

Work control in the work centers help to put manpower on the same page as per what needs to be done every time maintenance actions are required. The productivity levels required mean that not all manpower is unleashed on every work activity but some are reserved for research purposes while others are engaged in information gathering from ongoing work.

Inventory control requires that materials in warehousing be controlled to minimise misuse that could cost the enterprise a fortune. Every material of inventory such as spare parts, consumables, and tools in connection with maintenance activities must be released strictly based on maintenance work others and on no account should materials be released without documentation.

Cost control of manpower labour hours, materials and overhead is key to an effective maintenance program. Cost control impacts on assets and eventually rubs off on the total cost of ownership (TCO) carried in the books. Dependable maintenance is that which takes judicious steps to limit cost and possibly has in built mechanisms for limiting cost overruns.

Quality control that is the aim of maintenance activities is achievable if operations embarked upon use quality spares and manpower is trained in quality methods and deeply understands critical maintenance routines. Quality control is intended to deliver a safe working environment where quality maintenance and finished products can be delivered and therefore the realm of quality traverses the thinking, the methods used, and the handling of maintenance work.

Standardising Maintenance Management Around ISO55000 Compliance

Maintenance management being known globally as a requirement for best in class organisations is now standardised around ISO55000 compliance. The standard was launched in January 2014 to provide an overview of asset management, its principles and terminology, and the expected benefits from adopting asset management. ISO55001 followed in February 2014 by specifying the requirements for an integrated, effective management system for asset management. Not done, ISO55002 was released almost immediately to provide guidelines for ISO55001, but a revised and extended version of ISO55002 was released in 2018 to provide expanded detailed guidance to every clause of the ISO55001 requirements document, and clarification of the contribution of each requirement to the four 'fundamentals' of asset management: Value, Alignment, Leadership and Assurance. In September 2019 ISO55010 was released. This is a guideline enabling organizations to better understand why and how alignment between financial and non-financial functions are important in realizing value from assets.

Digital Maintenance Management

Computerised maintenance management is able to with data determine the delicate balance between selections of corrective breakdown repairs, preventive maintenance, or predictive maintenance approaches as budgetary directives may demand improved enterprise resource allocation. The data based maintenance helps to monitor pending, ongoing, and completed maintenance operations. Data maintained about machines helps to determine skillset pool requirements, tool inventory requirements, and spare parts requirements to deliver maintenance targets.

With more information available to help maintenance manpower do enough maintenance jobs there are possibilities for greater readiness of machines and work centers and therefore readiness for scheduling production orders, ultimately able to do contract jobs as more manufacturing facilities outsource jobs due to limited capacity constraints.

Central to digital maintenance management is a digital asset register that comprises attributes of all machines, facilities, and equipment used for manufacturing activities. These data are used to monitor the state of these assets and the types of maintenance to be executed on these assets offering the enterprise an optimised asset portfolio for productive revenue cycle engagement.

Digital maintenance management systems act as decision support systems (DSS) to management, to help management make informed decision for improved resource allocation.

Making Maintenance Work

While on the look out for machine limitations, relevant maintenance activities could be embarked on and recorded such as lubricating, cleaning, or adjusting and calibrating machines. Inspecting equipment to ensure proper operation and safety. Replacing parts that show deterioration. Checking, testing, and maintaining safety equipment, such as safety parameters, fire extinguishers, or alarm systems.

Four key elements make maintenance work outstanding: maintenance work order management, empowering with technology, gaining asset and maintenance intelligence, and transitioning from a reactive to proactive maintenance approach to guarantee the reliability of enterprise asset portfolio.

Accounting For Production

Production is an asset intensive activity. Embarking on production encumbers and consumes resources such as materials, manpower, and factory overheads in the form of machines and equipment that forms the capacity outlay in the transformation of materials to deliver finished products. The accounting that tracks and traces the value addition activities of a production enterprise are conversion cycle transactions because the asset mix in the books such as materials, labour, overhead are converted to another form of asset placed in the books as finished goods inventory or finished products.

Assessing the Performance of Production Activities

The performance of a production activity is ultimately judged on how well it contributes to achieving enterprise finance goals. This can be measured by the assessment of fiscal parameters such as the following:

Balance Sheet at Start - a statement of financial position as an inventory of assets and liabilities before the year's operations indicates the financial status of the enterprise;

Efficiency - the profit that is earned from the invested capital of the owners and externally borrowed funds measured as return on capital;

Liquidity - net cash flow representing cash that is available each year to pay all the bills of the expenditure cycle, including repaying borrowed capital and paying interest on loans when they are due;

Wealth - growth, net worth or equity. This is added to the owner's capital after all debts have been repaid;

Balance Sheet at End - a statement of assets and liabilities after the year's operations show the enterprise financial position as per how much the enterprise is now worth.

It is important to note that using these measures to assess manufacturing fiscal performance requires the use of management accounts.

The many goals of owners include the aim of having choices: about standard of living, what to work at, how much and how hard to work, ways to build esteem among peers and the local community, how well and where to retire and so on. The extent and nature of choices is heavily influenced by business wealth generation. The resources available influence the goal of the owners and how well the managers deal with variability. The more efficient and innovative the marketplace perception, the more net cash flow and wealth is created. Manufacturing business analysis measures efficiency of available resource use, cash earned relative to cash demand, and increases in wealth over time.

Contrasting Accounting For Merchandising And Services Enterprises

Merchandising and service enterprise accounting are similar in comparison to manufacturing accounting. These entities have straightforward income statement presentation around expenses incurred. With merchandising operations revenues are considered in terms of Sales and cost of goods sold (COGS). These are taken into account as a major expense driver to be considered separately in matching sales revenues, leading to the realisation of Gross Profit, which actually is derived from the revenue realised from merchandising operations. Merchandising organisations may hold a type of inventory known as merchandise that are acquired by engaging the expenditure cycle to support their merchandising activities, but those are accounted as sales revenue. Then the expenses in a merchandising enterprise is also applied as operating expenses against the gross profit derived because the cost of goods consumed within the period is excluded as this is reliably captured with COGS acting on sales revenue. Therefore, the income statement equation for a merchandising operation is as follows:

$Sales - COGS = Gross\ Profit - \Sigma(Opr\ Expenses)$

$= Net\ Income$

The revenue activities of a service business involve providing services to customers or clients as in some cases. On the income statement for a service business, the income from services are reported as revenues or fees earned. The operating expenses incurred in providing services are subtracted from the revenues earned to arrive at operating income. Service organisations may hold a type of inventory known as consumables to support their service operations, such as printer toners and papers, but those are accounted as expenses. In consideration of taxation, taxes could be added as an expense item on the list of expenses incurred or list of operating expenses. Nevertheless, the income statement equation for a service enterprise is as illustrated below:

$$\Sigma(Revenues) - \Sigma(Opr\ Expenses) = Opr\ Income$$

Accounting for Manufacturing Operations

Accounting for manufacturing operations follow a similar format to merchandising operations. But, unlike merchandisers, manufacturing is activity based and manufacturers produce the finished goods that they sell to customers rather than first engage the expenditure cycle with acquisition of merchandise to purchase those finished goods into inventory and then sell at a margin like the merchandisers do, the difference here is between merchandise inventory and finished goods inventory.

Significantly, manufactures account for raw materials as assets held in storage up till when they are transformed and released to storage after value addition as finished goods inventory with an intermediate holding account known as Work-in-Process (WIP) Account. Manufacturing production is initiated with a document described as a Production Order with attachments of Bill of Materials (BOM) itemising materials authorised to be consumed in the production of the specified units of products to be manufactured and a Routing Statement describing the processes, methods, and labour requirements for the production activity with an estimate of time and cost of materials and machines to be consumed. To commence production, a material requisition is raised, approved, and authorised to warehouse for the issue of the materials specified in them to be issued to production.

Accounts recognises the commencement of production in the books by debiting WIP account and crediting the different accounts of materials, labour, and overhead. At completion of production the Finished Goods Inventory (FGI) account is debited to increase its on-hand numbers, while the value addition process recognises those assets that were depleted to make the finished goods by crediting the WIP account to extinguish the earlier debit made at commencement of production.

Debiting the asset account of FGI indicate the emergence of something new which signifies that the manufacturer has actually delivered something new through its deftness of enterprise as Finished Goods which will show up in revenue cycles rather than acquire merchandise through the supply chain and the expenditure cycle as the merchandisers do. Significantly, WIP account is credited to Back-flush the materials, labour, and overhead that were earlier encumbered as debited when production was initiated and the reversal of the WIP account signifies the successful completion of production.

Where production is not completed within the reporting period that manufacturing was initiated what is stuck in production is reported as a WIP value asset in the Income Statement. Rather than account for cost of goods sold as it is with the merchants, manufacturers account for the cost of direct materials, direct labour, as well as overhead that make it into production as composite cost of goods manufactured, these prime costs are the major costs that differentiate manufacturers from merchandisers and there could be several cost elements that contribute to the Cost of Goods Manufactured (COGM) for the period to be subtracted from Revenues in arriving at Gross Margin for the period as shown in the income statement equation for manufacturers as follows:

$Revenues - \Sigma(COGM) = Gross\ Margin - \Sigma(Opr\ Expenses)$

$= Net\ Income$

Actual revenue earned by the manufacturer is rather considered as Gross Margin as opposed to the merchandiser's Gross Profit because the manufacturer is rather involved in value addition as opposed to a mark-up of purchases as done by the merchandiser and so stating his profit in terms of Gross Margin is consistent with terminologies of value chain analysis based on primary activities and support activities of the manufacturer in achieving competitive advantage in manufacturing. As with merchandisers, the sum of Operating Expenses when subtracted from the Gross Margin value yields Net Income for the manufacturing enterprise as indicated in the manufacturers income statement equation illustrated above.

Accounting for Materials

Material accounting in manufacturing recognises three kinds of manufacturing inventories being raw materials inventories, work-in-process (WIP) inventories, and finished goods inventories, as components used in production for transformation of component inputs to finished goods.

Accounting for materials first recognises a beginning debit balance in the raw materials account, subsequently, purchase of additional raw materials are debited to the raw materials account, which increases its debit balance. If at inspection, some of the materials are found to be defective they are returned to the supplier as return outwards but credited to the raw materials account, thereby reducing its debit balance.

When materials are issued to production for transformation and value addition they are recorded on the credit side of the raw materials account, reducing its debit balance. If at the end of the period production does not use up all of the materials issued to them they are returned back to warehouse and reported in the raw materials account on the debit side thereby increasing the debit balance of the raw materials account by the amount returned.

For inventory recoding, two systems are used to account for materials. The first is the periodic inventory system, while the second system is the perpetual inventory system. The periodic inventory is a classic inventory system where the inventory level is reviewed at regular intervals such as end of month to determine the COGM, whereupon the decision is made as to how much to order to bring the inventory level up to a given prescribed amount. The periodic inventory system is beneficial to enterprises as it allows a business to track its beginning inventory and ending inventory within an accounting period.

However, the perpetual inventory system records the movement of inventory continuously based on material in and out transaction source documents using digital technology to track inventory in real time with updates sent electronically to central databases.

Perpetual inventory systems provide a highly detailed view of changes in inventory with immediate reporting of the amount of inventory in stock, and accurately reflect on hand quantity levels.

In accounting for materials in production there are five classes of journal entries involved, these are as listed in the table below:

	ACTIVITY	JOURNAL POSTING
1	*Purchase of materials on account*	**Debit** Raw Materials Inventory **Credit** accounts Payable **Credit** Vendor account
2	*Return of inferior materials to supplier*	**Debit** Accounts Payable **Debit** Vendor account **Credit** Raw Materials Inventory
3	*Issuance of direct and indirect materials to*	**Debit** Work-in-Process Inventory **Debit** Manufacturing Overhead **Credit** Raw Materials Inventory

	production	
4	*Return of excess direct and indirect materials to warehouse*	**Debit** Raw Materials Inventory **Credit** Manufacturing Overhead **Credit** Work-in-Process Inventory
5	*Payment for materials to vendors*	**Debit** accounts Payable **Debit** Vendor account **Credit** Cash

Accounting for Labour

Manufacturing labour is the engagement of manpower that exerts physical and mental effort for the transformation of raw materials to finished goods. Accounting for labour includes salaries earned by factory employees, such as basic salaries, overtime pay, medical benefits, and bonuses for exemplary performances. Factory workers may also earn daily rates, overtime rates, special pay, and night differentials when they work night shift to engage production resources round the clock to bring maximum benefit to the enterprise in production exceptions. An enterprise uses payroll as a tool for computation of what employees earn by providing clock cards that are used by workers to clock in their arrival time and clock out their departure from factory premises to build up the time register, the clock cards could also be used as a form of identification and for access control within the premises in some cases. At every pay cycle the supervisor collates the hours worked by each employee from the time register to derive standard hours, overtime hours, and night hours and applies their respective hourly rates to determine a spreadsheet of how much is due to each employee for the period through a payroll, and advise individual employees with a payslip detailing their earnings, where legitimate deductions are made to derive their net pay which is what is eventually paid to the employee. The journalising of payroll is carried out as follows:

ACTIVITY	JOURNAL POSTING
Charge factory payroll to production	**Debit** Work-in-Process Inventory **Debit** Manufacturing Overhead **Credit** Factory Salaries And Wages

Accounting for Manufacturing Overhead

Manufacturing overhead include all costs in the production process that are not direct materials or direct labour not directly traceable to the products completed but are still necessary to be incurred to transform raw materials into finished goods.

Manufacturing overhead include indirect materials, indirect labour, factory insurance, factory depreciation of equipment and machines, repairs and maintenance of factory asset portfolio, factory rent, and factory utilities.

Service or support departments such as purchasing, personnel, manufacturing, and maintenance departments support the production process and costs incurred in them should be allocated to the production department for determination of production costs of a product since they contribute to the product in an indirect way.

Overhead expenses can be divided into three general categories: company overhead, selling overhead, and administrative overhead. These expenses cannot be directly linked with manufacturing products or providing services. Manufacturing overhead is calculated by adding all the indirect factory-related expenses incurred in manufacturing a product. This includes the costs of indirect materials, indirect labor, machine repairs, depreciation, factory supplies, insurance, electricity and more.

The manufacturing overhead account is a holding account for the actual overhead costs incurred (debits) and applied to work-in-process (credits). Actual overhead costs flow into the WIP account as they are incurred. Applied overhead costs flow out of the account as the jobs proceed through the production process.

The direct method is the most widely used method for allocating overhead where it allocates each service department's total costs directly to the production departments. The journal entry to apply or assign overhead to the jobs would be to move the cost from overhead to work-in-process inventory. The most common method for disposing of the balance in Manufacturing Overhead is to make a direct adjustment to COGM.

Expenses normally have a debit balance, and the manufacturing overhead account is debited when expenses are incurred to recognize the incurrence. When the expenses except overhead costs such as general administrative expenses and marketing costs are allocated to the asset, the work-in-process inventory account, the expense account manufacturing overhead is credited.

Closing a manufacturing overhead account demands determining if the amount involved is substantial. If the amount is inconsequential, then the balance is closed directly to the cost of goods sold account. If the amount is material, the amount is allocated

Support Activities	Firm Infrastructure					Margin
	Human Resources					
	Technology Development					
	Procurement					
	Inbound Logistics	Operations	Outbound Logistics	Marketing and Sales	Service	

Primary Activities

between the work in process, finished goods and the cost of goods sold accounts.

Value Chain Analysis

Manufacturers report their income statements by subtracting COGM from revenues to arrive at gross margin which when operating expenses is subtracted from it gives the income for the period reported. The margin seeking approach in manufacturing is a process known as Value chain analysis (VCA) where manufacturers identify their value addition primary and support activities that impact its final product and then analyze these activities with an aim to reduce costs or increase differentiation. Value chain represents the internal activities a manufacturer engages, such as production, marketing, and the provision of after sales service, when transforming inputs into outputs in its value addition where the intensity of the manufacturers margin depends on how deep it penetrates the value chain with primary and support activities inherent in those value chains.

The purpose of value-chain analysis is to increase production efficiency so that a manufacturer can deliver maximum value with the least possible cost. Cost overrun is an issue in manufacturing and manufacturers that get it right do so with an uncanny ability for cost reduction. This involves all of a product's stages of development, from its design, to procurement strategies and intermediate inputs, its marketing, its distribution, and its customer support services. The value chain concept has several dimensions. The first is its flow, also called its input-output structure.

VCA can expose strategic and operational misalignments within chains, be it the demand chain or the supply chain, and the consequential misallocation of resources, with the end product in view and hence opportunities for improvements which create value and economic sustainability. The outcome being that such analysis results in identifying the activities that are a source of cost or lead to a differentiation advantage and those that could improve the competitive advantage of the company or of the product to secure marketplace product viability.

Product costs are recorded as an asset on the balance sheet until the products pass through the revenue cycle after they have been sold, at which point the costs have passed through the expenditure cycle and are recorded as an expense on the income statement where income from the revenue cycle and expenses from the expenditure cycle are brought together for the period for comparison and identification of a profit or a loss for the period based on realised margin. The income statement indicates how the top line revenues are transformed into net income, net profit or net loss based on the difference between the sum of income and sum of expenses to show stakeholders whether the company made money or lost money during the period reported. In just totalling revenues and subtracting expenses to find the bottom line companies establish the reality of the value chain as it applies to realize margins for the period based on a single step income statement. However, the multi step income statement goes further to derive the realized margin as gross profit, then calculating operating expenses which when deducted from gross profit yields operating income before taxes, while the final step is to derive the net income after tax deductions. The income statement portrays the cash generation capacity of the enterprise, the penetration of its value chain, and the viability of the primary and support activities the enterprise is engaged in.

Make-to-Order Manufacturing Accounting

Mate to order manufacturing accounting moves digital marketing, sales channel, fulfilment, and billing processes of the CRM subsystem into the revenue cycle with a customer order and lets the value addition process of the PRM place the mechanisms of the conversion cycle at the disposal of finished product order customisation realisation towards customer fulfilment steps. Ultimately, the delivery of the finished product results in journalising of sales and inventory, which when posted, the sales ledger control account is debited, while the sales account is credited. When delivery is made the COGS account is debited, the customer's account in the subsidiary ledger is debited with the customer's posted invoice as pending and inventory account is credited. Finally, when the customer makes payment, cash account is debited while the sales ledger control is credited and the customer's account in the subsidiary ledger is equally credited with the same amount as well.

Make-to-Stock Manufacturing Accounting

With make to stock accounting it is rather the VRM subsystem rather than the CRM subsystem that moves the replenishment process through an MRP breakout to layout material requisition internally which triggers the manufacturing accounting system to encumber materials to make certain products considered as sub assemblies in the manufacturing of other finished products or simply to be showcased in enterprise showrooms for customer off the shelf purchases or vendor volume purchases for the supply chain.

It is first to recognise the purchase of materials on account with journal entries of a debit to the raw materials inventory account, then a credit to accounts payable and a credit to the vendor's account. Subsequently, materials are issued to production and recognised as debit to WIP inventory account, a debit to manufacturing overhead, and a credit to raw materials inventory. Eventually the vendors are paid with journal entries of debit to accounts payable, debit to vendor's account, and credit to cash.

When the finished products are released to inventory the finished goods inventory is debited and the WIP account is credited. From the finished goods inventory the enterprise now has merchandise for sale to customers

Activity Based Costing

Activity based costing (ABC) identifies activities in an enterprise and assigns the cost of each activity to all products and services according to established cost drivers, being the actual consumption rate by the resources consuming them, by focusing on activities and cost drivers. The goal of ABC is to accurately allocate costs to resources.

ABC is most helpful in manufacturing formations, where overhead costs constitute a large portion of production cost. Activity-based costing is especially useful to allocate indirect costs to items that are difficult to track and assign. The main benefit is more accurate product overhead costing. ABC provides realistic costs of manufacturing for specific products by allocating manufacturing overhead more accurately to products and processes that consume those activities specified and also identifies inefficient processes and targets them for improvement. This way ABC determines product profit margins more precisely.

ABC assigns manufacturing overhead costs to products in a more logical manner than the traditional approach of simply allocating costs on the basis of machine hours. Activity based costing first assigns costs to the activities that are the real cause of the overhead. It then assigns the cost of those activities only to the products that actually demand the activities. Manufacturers use activity based costing when overhead costs make up a significant percentage of overall expenses. Manufacturers also use ABC when they produce product lines of varying quantity and complexity or produce a broad array of products requiring various service support levels.

The components of ABC are resources, activities, cost objects, and drivers.

Resources such as people, facilities, and costs associated with people and expenses are the economic elements consumed while performing activities. They are the core elements that let enterprises operate.

Activities consume resources and drive costs to cost objects. The lowest-level definition of what is done; activities are the foundation for measuring activity costs.

Cost objects represent cost information grouped by profitability dimensions, such as products, customers, and channels. With resources and activities linked to cost objects, which are the final results of the activities performed, they are often the focal point of profitability analysis.

Drivers are transactional, duration, and intensity assigns to monetary amounts from one object to another throughout the model (calculated by amount, percentage, spread even, and direct) in different ways depending on assignment type and object type, even across business units.

In ABC costs in cost pools are allocated to resources that consume them based on established cost drivers that simulates the rate of consumption of the resources from the cost pools to deplete the pool of costs.

Financial Resources Management Systems In Manufacturing

Digital financial resources management (FRM) structures and records fulfilment, procurement, and value addition transactions with double entry self-balancing postings to the general and subsidiary ledgers in real time to sustain a robust fiscal rendering system. Ledger definitions are initiated with a customized framework of industry specific chart of accounts provided specifically for manufacturing enterprises being somewhat different from what is obtainable for service and merchandising concerns.

An FRM system provides options to record settings that coordinate necessary postings in control accounts as well as set budget ceilings on expenses for administrative control and automation of receivables and payables.

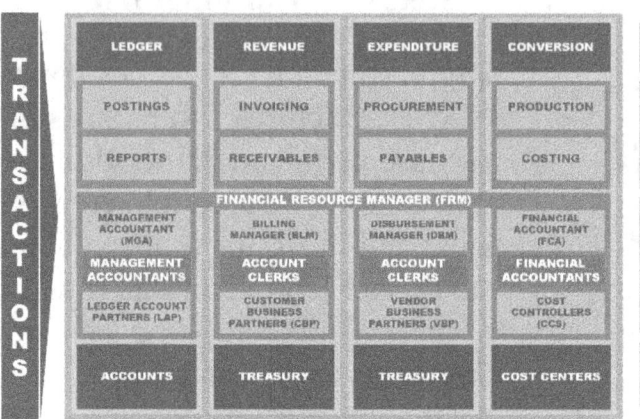

ACCOUNTS

FRM further ensures that receivables, payables, and appropriations are well accounted for through ledger, revenue, expenditure, and conversion cycles with four financial statements; income statement shows how the company is doing, statement of owners equity shows what is done with profit, balance sheet show assets, liabilities, and capital values, while statement of cash flows show where the cash came from and what were done with them.

The FRM use case allocates roles to five talents of treasurer, budget accountant, cost accountant, and tax accountant, under the leadership of the Chief Financial Officer (CFO) the roles have various posting responsibilities, however smart journals post from source documents generated from the CRM,

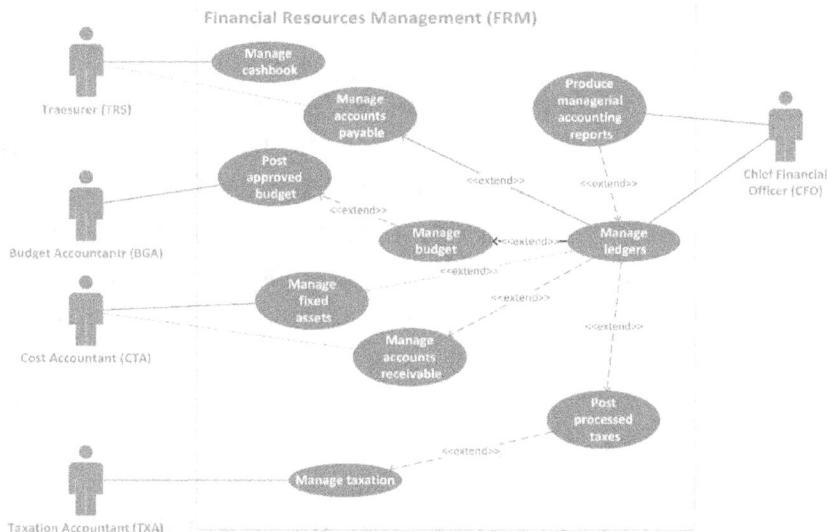

VRM, and PRM modules automatically to ensure proper coding for posting, classification, and summarising to meet the requirements of the ledger cycle.

Ledger Cycle Operations

FRM places all accounts in a ledger cycle of five phases of recording, classifying, summarizing, reporting, and interpreting that are necessarily carried out every period to make meaning out of accounts that are collected in ledgers and ensure these ledgers are updated and suitably maintained according to prevailing guidelines.

Recordings are carried out to chronologically journalise transaction sequences in order of occurrence for sales, purchases, acquisitions, expenses, disbursements, borrowings, receipts, bad debts, and depreciations in smart journal entries from executed source documents containing details of business transactions, such as receipts, customer and vendor invoices, purchase orders, cheques, deposit slips, and vouchers, then the source documents are archived away.

The classifying phase through intelligent parameter settings make the provided smart journals able to class, order, and categorise posted transactions according to reporting expectations and close out end to end double entry postings and self balance all posted accounts, drawing a new running balance on each account based on its previous balance.

The summarizing phase consolidates totals brought about by data transformation and aggregation of general ledger and subsidiary ledger sub classifications. The reporting through rendering of financial and managerial reports are able to help management guide their operations. Finally, the interpretation phase determine margin profitability, cash liquidity, and obligatory solvency of the business.

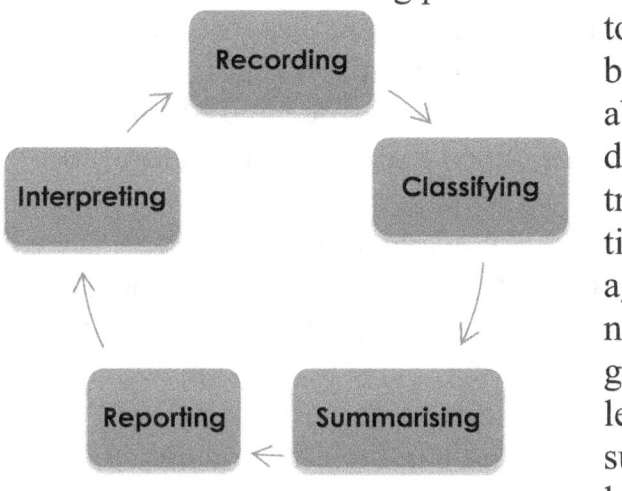

All enterprise accounts are to be held in ledgers. After adjusting journal entries (AJEs) are entered to these accounts through revenues and expenses accruals that recognise revenues earned and expenses incurred, deferrals that places recognition of expenses made to a future date, depreciation that recognises the expense of asset deterioration, bad debts that recognises a revenue collection loss, and correction for errors, it is safe to prepare management reports such as income statements as statement of financial performance and balance sheets as statements of financial position from the accounts for further analysis after AJEs have helped to update the accounts. FRM accounts in the ledgers are cyclically posted from smart journals that know exactly how the debit and credit sides are to be posted due to their intelligent parameterisation, and maintain a running balance of each account so that a new running balance is established after every single posting to each account in both the general and subsidiary ledgers. Journals are smart to the extent that they can differentiate between debit and credit balance accounts such as assets and expenses as debit balances, while liabilities, capitals, and revenues are credit balances, and their posting characteristics based on control account settings. Industry standard chart of accounts such as for merchandising, manufacturing, and service businesses are provided so that the enterprise can have standard reports as fiscal aggregation and rendering of these accounts.

Most importantly, at period end the ledger cycle is closed by closing credit balances in all revenue accounts to the income summary account, setting balances of all revenue accounts to zero. Next, debit balances in expense accounts are closed to the income summary account, setting balances of all expense accounts to zero by posting a credit to each of them with the exact same amount in each expense account debit balance. Also, income summary account is closed to the retained earnings account in case of corporations or to capital accounts in case of sole proprietorships. Finally, dividends accounts are closed to retained earnings account. Thus bringing a period in the ledger cycle to a close. Prior to this, it is necessary to post prepayments, accumulated depreciation, and prepaid supplies as stated above so as to give a true picture of the ledger cycle in view of assets employed.

With the self-balancing characteristics of FRM ledgers, the enterprise will always have numerically balanced trial balances. Very little work is required to determine the balance sheet, the income statement, and cash flow statements. The ledger cycle helps to pace out the accounting process by breaking down operations into the phases of recording, classifying, summarising, reporting, and interpreting that make the accounting cycle a most welcoming and flourishing process for enterprise operations, rather than the drudgery of traditional manual accounting.

Revenue Cycle Operations

With an upbeat fulfillment process enterprises will be faced with a stack of receivables arising from its billing operations. Ultimately, enterprises will want to push for all their customer engaged hours and customer deliveries to be billable.

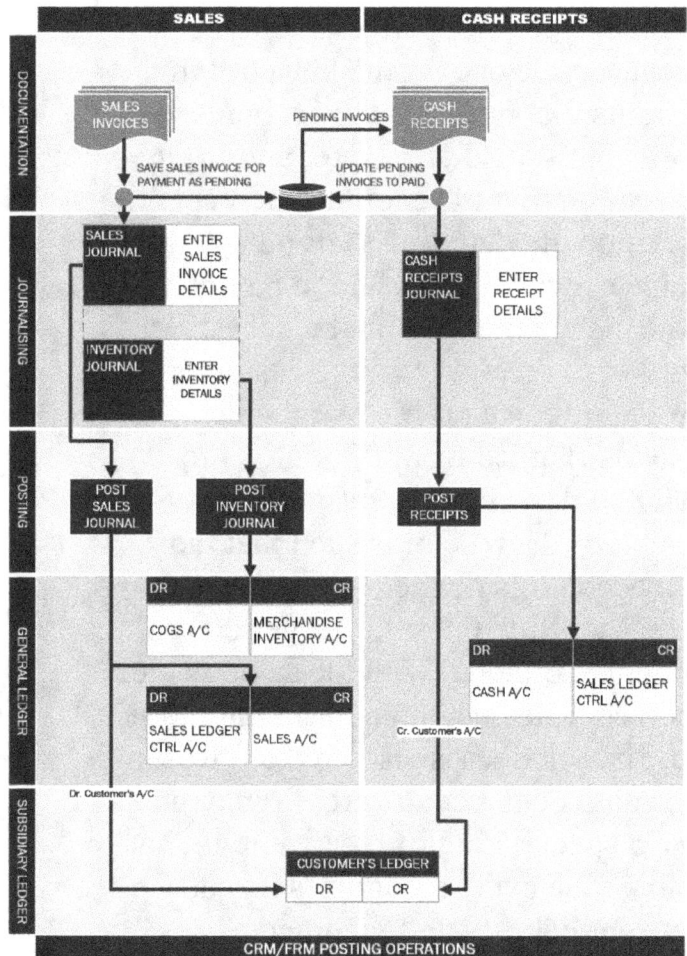

The revenue cycle works the electronic parameters that define receivables from the settings given each customer account for prepaid, cash on delivery, or on account based on the relationship goals of the enterprise with the customer. The first task in the revenue cycle is credit approval, based on which the customer order is filled and deliveries are ordered. After which invoices follow deliveries made. These invoices are filled off as pending invoices and have their respective due dates that kick the revenue cycle into collection mode. The revenue cycle checks the balances in the accounts that make up the accounts receivable subsidiary ledger and triggers the sending out of account statements to all customers detailing their outstanding balances, which then send in payments to clear such outstanding balances. For every cash receipt from the customer, the pending invoice on file from which it is raised is flagged from PENDING to PAID, and the double entry process for posting of the receipt is such that the cash account is debited so its balance increases while the sales ledger control account is credited with the same value received. The customers' account in the subsidiary ledger is also credited tending to bring its balance to zero. Prepayments pose a special kind of receivable in that the revenue cycle recognizes its entry as a current liability and recorded as unearned income, and progressively deducts the applicable amount per revenue cycle as payment received from the customer and subsequently updates the customer account to reflect this while entering a ledger posting of debiting cash account and crediting the customer account while also debiting the liability account and crediting accounts receivable control account.

Expenditure Cycle Operations

The enterprise procurement process generates payable commitments due to vendors from its purchasing operations based on replenishment orders and sundry purchases. The expenditure cycle processes authorised payables based on the relationship criteria with vendors.

The first task in the expenditure cycle is approval of vouchers comprising vendor invoices, requisition orders, and Receiving Reports based on which the vendor invoice is approved after certifying a connecting matching criteria across these documents. These invoices with their Receiving Reports were filled off as pending invoices and have their respective due dates that kick the expenditure cycle into disbursement mode.

The expenditure cycle checks the balances in the accounts that make up the accounts payable subsidiary ledger and flags outstanding balances for payment and sends out remittance advise to all vendors with outstanding balances, to which the enterprise send accompanying payments to clear their outstanding balances in their favour.

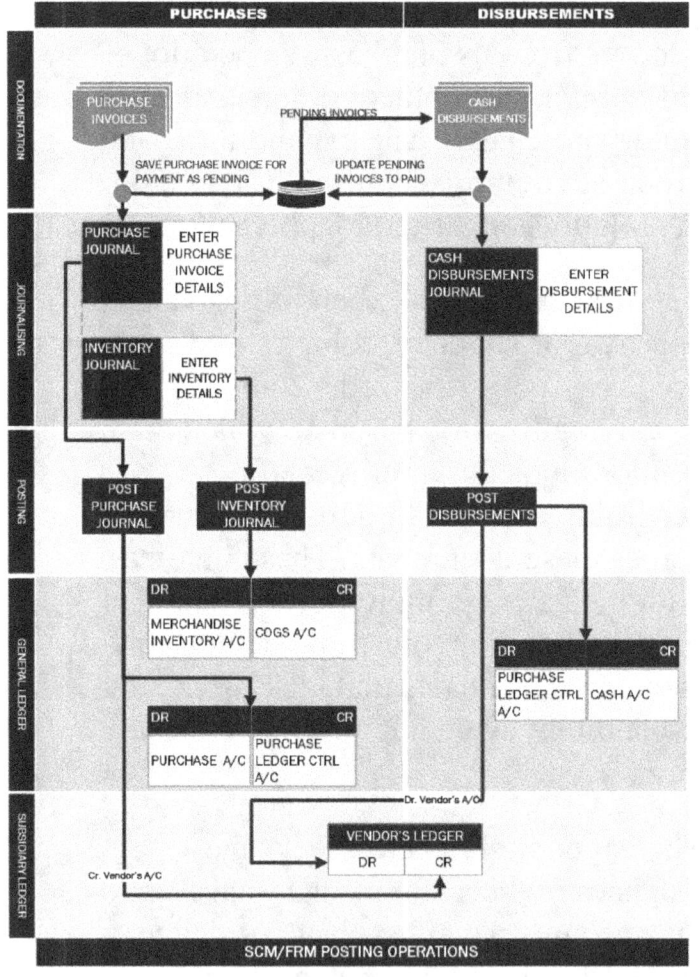

For every cash disbursement that is made out, the pending invoice on file from which it is raised is flagged from PENDING to PAID, and the double entry process for the posting of the disbursement is such that the purchase ledger control account is debited with same value disbursed, while the cash account is credited so its balance decreases, the vendor accounts in the subsidiary ledger are also debited tending to bring these individual balances to zero. Prepaid expenses form a major part of the expenditure cycle. There are two parts to this. The first is the entry made at incursion of the prepaid expense when the prepaid account is debited and the cash account is credited. Then periodically to recognise the expense the entries are to debit the individual expense concerned and credit the prepaid expense account progressively until the prepaid expense account is wiped out.

Because of the adverse effect of expenses on liquidity organisations often cap expenses with budget ceilings for every period so that the workforce goes about their activities with the general awareness of a ceiling on the possibility of introduction of additional expenses to the company's operation. This calls for frequent budget proposals and being inputted into FRM for budgetary controls whereby these guidelines limit expense to the barest minimum requiring administrative approvals and authorisations to admit budget override once it gets close to predetermined ceilings.

Just as budgeting helps to make more liquidity available to the enterprise, there must also be clear policy guidelines to steer the company clear of liabilities, as liabilities on their own are capable of generating future expenses. However, statutory liabilities portend major issues to every company as their non-recognition could plunge companies into long drawn legal tussles with government. Key among these is tax liabilities and levies. They must be negotiated and booked early for payment to avoid operational hardships being brought on the company by agents of state.

Conversion Cycle Operations

The conversion cycle's value creation process engages warehouse materials, skilled labour, and the production arm of the business, spanning the ERP modules of FRM, PRM, and WRM.

Operations of the warehouse arm is for the costing of materials released, while operation of the production arm is for costing inputs into the value addition process of

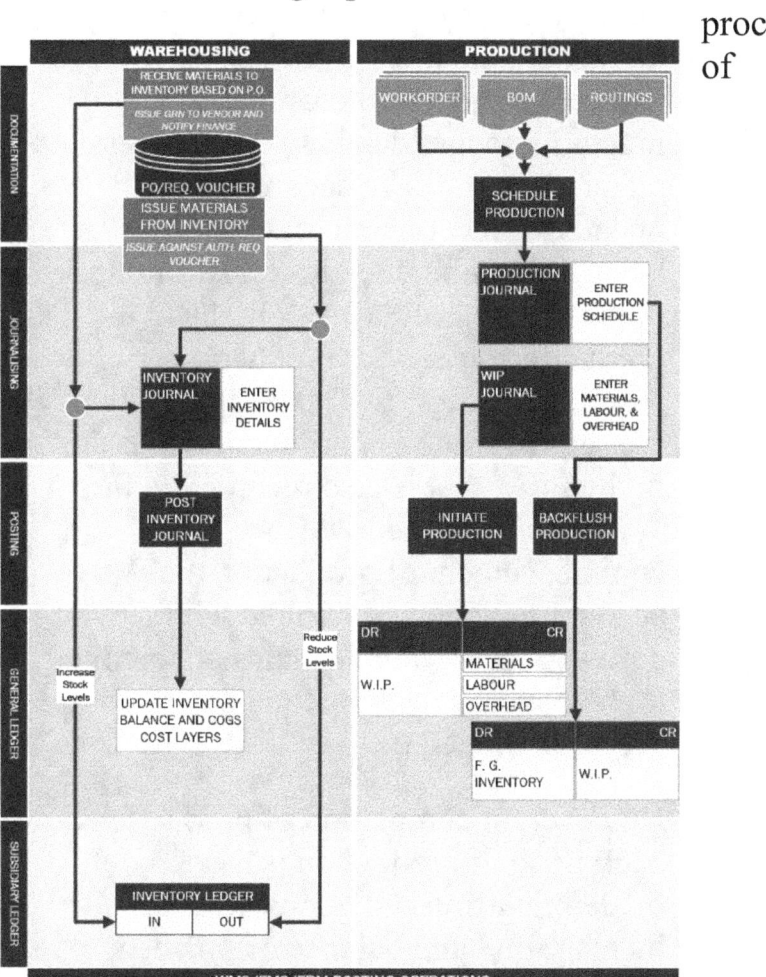

production. Valuation of materials commences from when they are received into inventory through a Receiving

Order by the receiving unit of the warehouse.

The FRM conversion cycle uses a number inventory costing techniques to do this by updating inventory balances and cost of goods sold cost layers in the inventory ledger database where every material item has an account that is updated at receipt and at issue and a balance is maintained to reflect the stack of prevailing cost layers. On the production side, after production is scheduled with WO, BOM, and routings are sequenced to the production journal in the PRM that is also recorded to the WIP journal in the FRM. At commencement of production the WIP account in the GL is debited while corresponding WIP accounts are credited to the materials, labour, and overhead accounts in the GL.

However, at completion of production a back flush of production is carried out by debiting finished goods inventory to increase on hand balance of finished goods account and crediting the WIP account somewhat reversing the earlier balance in the WIP account that was recorded at commencement of production. The conversion cycle prominently features cost centers that are special accounting structures for cost valuations.

Infrastructure For Production

Accessing production requires enterprise infrastructural outlay to accommodate related factors of machines, manpower, margins, materials, and methods, and give significance to plans and schedules of manufacturing. Infrastructures are essential to support production.

Production processes such as BOM formulation, routing composition, work scheduling, and production execution are supported by specific infrastructures in the form of logistics, work centers, plan boards, and shop floors.

Enterprise infrastructures are required to speed up the growth of business. Such basic systems and services help production function properly with the right priorities, with a global reach, and with the right manpower. These infrastructures provide the framework on which to build production requirement plans, the assets that generate materials that make it into the revenue cycle. Based on the capacity of infrastructure to facilitate production, no infrastructure is dispensable; every single device is critical for productivity and capable of shutting down production or become a bottleneck to production throughput if not adequately integrated to support production.

Production infrastructure is an instantiated form of devices for an enterprise in a local environment and a departure from public infrastructures or utilities. They eventually plug into the public infrastructures, but that is when the role of enterprise owned infrastructure have played their part.

With the advent of Industry 4.0, the production capabilities of industries have become connected, elevated, scalable, and smart, all thanks to the performance of getting production infrastructures right.

Logistics Infrastructure

Material handling is a major issue in manufacturing. Materials meant for transformation into finished products need to be acquired from the supply chain, value retained in the warehouse, subjected to value addition processes in the conversion cycle, then moved into finished goods inventory at completion of the conversion cycle.

Material handling is key to presenting materials at various production stops for value addition until finishing point of the material transformation. Materials need to be handled for grading, movement, and storage with appropriate information flows to alert processes and mechanisms of the state of material availability and expected job, which will also result in greater manufacturing performance. As more and more materials must make it into manufacturing so also does bigger logistics infrastructural outlay to realise material objectives. The scale of material handling performance determines the scale of finished products delivered in manufacturing establishments, optimising production by connecting warehouses to marketplaces, shop floors and work centers.

The outlay of enterprise inbound logistics with vendors on the supply chain to derive optimal value from the procurement channel, and outbound logistics with customers of the demand chain is the fulcrum on which the conversion cycle's production efficiency is derived and when forward logistics and reverse logistics are taken into account then there is room for industrial growth, transformation, and environmental sustainability.

Smart logistics infrastructure is the fuel for industrial growth and every industrial sector depends widely on far-reaching logistics outlay for fulfilment, sourcing, procurement, and replenishment. More enterprises are banking on adaptation of more technology and IT solutions to support logistics infrastructures.

Logistics primary activity is the transportation infrastructure for primary, secondary, and tertiary packages through transportation systems. Inbound logistics engages sourcing, procurement, and replenishment processes of the supply chain with preferred vendors for smart procurement operations from an enterprise instantiated supply chain to support a robust replenishment and production environment. Focus of logistics is on the interplay of orders, materials, and vendors being critical to adopting the purchase channel and its logistics to ensure a beneficial value addition for the enterprise.

While on the flip side engaging outbound logistics provides options for despatch distribution mechanisms through fulfilment processes of the CRM to sustain a sales channel of satisfied customers with improved customer satisfaction numbers as customer expectations are for faster product deliveries for greater marketplace competiveness.

The scale of logistics infrastructure is dependent on the scale of warehousing where accumulation for value retention lays out various structured storage systems for value retention and enhancement to support procurement, fulfilment, and value addition activities.

Finally, forward logistics is what most enterprises do with inbound and outbound logistics, connecting demand and supply chains. But in todays world of material recycling, there is a new order, with reverse logistics that mops up waste materials from the environment for re-transformation into a new raw material, so that they can be represented to the value addition processes as fresh raw materials to derive new finished goods, thereby withdrawing their otherwise harmful effects from the environment as wastages viable as non bio-degradable materials.

Logistics is a disruption in the manufacturing environment as they are not readily feasible for all enterprises and as such most enterprises have to make do with third party logistics solutions as a tie back to their established demand and supply chains to maintain marketplace relevance to satisfy customer expectations and scale to vendor capacity. Strategic infrastructures pave the way for a more efficient logistics and manufacturing complex.

Material Handling

The efficient handling and storing of materials by moving, packing, and storing in suitable locations safeguard the proper commencement and completion of material based value addition processes. Material handling being a short-range movement of materials aims to present materials in the most suitable form through robust and safe handling, to storefronts, work centers, and warehouse storage systems, ensuring adequate value retention in all cases where materials are accumulated to meet required capacities.

Work Center Infrastructure

Work centers are work environments where methods and processes of production that are designed and composed into structured procedural formulations for the transformation of materials into finished products are implemented. Work centers bring together manpower groups and crews of the enterprise by delivering the tools and technologies to get people working as collaboratively as possible to deliver on tasks and targets of production by following through with the production routings handed down for the production of finished products.

Work centers are the cultivation grounds for production where planned work are scheduled to be executed under the governance of skilled manpower who team up to tear down the directives of provided production routings to consume materials in the supplied routings of documented methods to yield product expectations. Today's work centers are more sophisticated and technology driven to ensure connectivity with machines, manpower, materials, and methods for the best production mix and flawless yield delivery.

The capacity to replicate defined work centers makes it conceivable for work centers to be virtually delivered at will at any point on the globe to retool and instantiate any production requirement for consumption customised for various regions. Where replication is possible manufacturers are able to pursue globalisation strategies to the right margin, replication work centers to boost production capacity.

Work centers are specified on routing sheets as WIP stopovers to be worked on by machines, being transformed while still in the work-in-process stage. Materials may traverse just one work center or multiple work centers to complete their production sequence. When they arrive a work center they are worked on with specified machines in which case the work center is considered a machine center. Where skills are of paramount importance in the transformation of materials the routings indicate specific skillsets as applicable to take the materials through expected transformation in which case this time around the work center is considered a skill center.

Queuing of production to work centers ordinarily initiates a WIP transaction followed by material handling for transformation until production completes at which time machine and labour hours will have been consumed to deliver the right margin at which the product embarks on its journey to shelf spaces to derive its marketplace survival.

Production grade work centers must be connected, equipped, and able to deliver the specified yield assigned to them. Work centers therefore must also be safe and ergonomic for skilled manpower and take less time to get used to. In building the right work center enterprises must take the needs of factory workers into account so that manpower assigned are able to produce to the stated nameplate of the work center.

Plan Board Infrastructure

The plan board is a major infrastructure of manufacturing, as simple as its construction is, the plan board is where manufacturing horizons are sorted out and shuffled to meet expectations by allocating materials, machines, manpower, and work centers to pending and on-going production orders as action plans for the entire organisation. The plan board holds much for the enterprise as an advanced planning and scheduling tool coordinating the timetable on which production revolves from where production orders are given prominence to assume the production runway for execution. The plan board shows which work centers are doing what, and perhaps which manpower crews are doing what, based on the work centers scheduled to handle specified orders.

The plan board for production according to the sequence of the enterprise is divided into swim lanes with each swim lane specifying a work center and on each swim lane are rectangular boxes representing production orders with their length being an estimate of the number of days the production order is to be executed within the specified work center. In specific circumstances, the production orders indicate resource allocation, which warehouse must encumber for production release during planning.

In a digital world, the plan board is a production playbook where schedulers simulate drag and drop scenario production orders on a visual big screen for visual display and visual control helping production teams identify opportunities for improvement to create different scenarios for presentation before a particular scenario is accepted by stakeholders and benchmarked for future planning as well. Fully subscribed plan scenarios become part of the production facility's dashboard and its measures as part of production KPIs for tracking the productivity levels of the plant to grow the top line of the business.

Work prioritisation is a major plan board issue as enterprises must assign weights and priorities to every production order in advance and be able to manage priorities, complexities, and urgencies of orders throughout the lifecycle of order progression to its completion and handover to warehouse. Plan board flexibilities must address team concerns, contain erratic demands, and offer countermeasures to production pressures for an in depth problem solving approach to production planning.

In some manufacturing sites, the plan board is considered a Kanban Board, according to Japanese manufacturing strategies, for more efficient processes that ties in the supply chain, tracking orders, coordinating materials, and dispatching replenishments into work centers for on time delivery of production orders by tracking them through the sequences of pending, in progress, and completed orders.

Work center loading must also be considered when assigning orders to work centers to avoid downtime or bottlenecks in production.

Shop Floor Infrastructure

The shop floor is an exceptionally large industrial space where outputs of preliminary processing from work centers are taken for final assembly and production flow into the final product of production after all sub assemblies, kits, and components have been processed into forms required and suitable for final production in the work centers and moved into the shop floor for final production.

The shop floor is the main production area where production flows continuously into the final product and moved into the warehouse as they are completed. Power tools are used on shop floors because they operate at a higher scale as they absorb preliminary production of sub assemblies from work centers and process them into primary packages and subsequent secondary packaging as well and must be able to robustly handle materials and move materials through to completion at the highest rate possible being the highest level of production in the enterprise and for the greatest productivity the enterprise can afford.

Shop floors often employ advanced automated systems for material handling and operated by skilled manpower that must be appropriately trained to observe safety protocols to minimise work place injuries and scrap rates. The shop floor is an integrated equipment hall, inventory, and storage areas complete with manpower briefing area for crew based production briefings, presentations, and production reports.

Shop floor management is a major activity of every production enterprise where increased productivity and efficiency is desired. It also calls for greater control so that it does not spin out of proportion to mismanage materials. Quality and costs need to be managed for marketplace competitiveness. Manpower also need to be appropriately managed for the best human factors such as reduction of absenteeism, wastages, and injuries with functional attendance registers and injury show back with remedial action and rolling statistics of lost time injuries and man-hours lost, as well as current scrap rates. Shop floor manpower need firm supervision and rigid structure to achieve desired productivity. Real time data control of shop floor parameters is preferred for preventive and predictive maintenance settings for the shop floor equipment and machines.

Groomed Production Systems

Manufacturing is the mainstay of global economic sustainability, as enough products must go round for the world to go round. Hence, the world has historically groomed several manufacturing systems to sustain global appetite for the consumption and acquisition of essential products.

New manufacturing systems are derived through a strategic combination of a need, materials, financing, manpower, and enabling technology. As economies mature into new realms they deliver new manufacturing systems to sustain their economies.

The increasing demands for consumer goods require a framework for delivering sustainable manufacturing platforms. A manufacturing system is a collection of methods, processes, and mechanisms for producing desired products is a manufacturing system. They process materials within the system and value is added to materials as they progress from machine to machine. This value addition of the conversion cycle is on a scale of economy that flourishes enterprises engaged in them because they create value and enjoy favourable margins by delivering these values to customers who sustain the enterprise revenue cycle to help it legitimately keep the margin it has created in the process. Groomed manufacturing systems are an encapsulation of best practices through learning points to deliver best of breed production at a scale akin to results consistent with what is obtainable with stated outcomes from the recommended templates of the manufacturing system outlay. The teardown system approach offered by groomed manufacturing systems is a hands on for inductees of manufacturing best practices that help to systematize manufacturing and consistently deliver scalable results.

Mass Production

The Industrial Revolution in the period from about 1760 to sometime between 1820 and 1840 catalyzed by the emergence of capitalism, European imperialism, efforts to mine coal, and the effects of the agricultural revolution gave rise to mass production, happening in Great Britain, continental Europe, and the United States. Mass production enabled the manufacture of considerably large quantities of identical products using automation and assembly lines to manage repetitive, flow, and series production at low costs. Mass production became fascinating to the marketplace, and based on market demand, production became a continuous flow; work-in-progress in factories became limited as the shop floor absorbed all materials that were encumbered to be consumed in production because standardised production parameters released products of high quality. And with division of labour with manpower developing adequate workplace skills, supervision was easy with fewer instructions to a knowledgeable workforce. Material handling got a boost with machines handling materials at most times with less exertion of manpower effort. Queuing of materials to work centers was not necessary as all materials went straight to the shop floor and so the shop floor got prominent with materials getting on the shop floor with little preliminary processes and so the flow of materials became continuous.

In mass production, production lines are aided by assembly lines in a process in which interchangeable parts are added to a product in a sequential manner to create end products. The machinery and manpower consumed to produce items are stationery along the line and the product moves through the cycle, from start to finish.

Mass production benefited from the derivation of five economies of scales: technical economies of scale improved efficiency and size of production; purchasing economies of scale helped the enterprise buy in bulk through long term contracts so that the unit cost of materials became substantially reduced; managerial economies of scale increased the specialisation of managers so that the management structure were able to effectively run the business and tackle more managerial challenges robustly as they were specialists in their various fields of endeavour; financial economies of scale enabled enterprises obtain financial leverage at a low interest rate and provided access to a wide range of financial instruments; marketing economies of scale reduced advertising costs by spreading the cost of advertising over a wide range of media markets; and technological economies of scale took advantage of mechanisations in material handling to deliver more materials for processing to a shop floor run by motorised assembly lines and conveyor belts.

Mass production had many positive effects. It raised the standard of living, increased wealth, increased production of goods massively, access to healthier diets, better housing, cheaper goods, and education increased as a factor of the industrial revolution. With mass production productivity increased, products became uniform, goods assumed lower costs, quality of life became higher, production became faster, there were less errors in production, division of labour made way for job specialisation, increased worker safety, and rapid evolution of manufacturing.

Toyota Production System (LEAN Production)

The Toyota Production System (TPS) are a set of concepts and methods developed and implemented by Taiichi Ohno, the Chief Engineer of Toyota between 1940 and 1950, for the production of Toyota cars, with a basic philosophy for efficiency in manufacturing by cost reduction based on thorough elimination of waste. The TPS is a production system that systematizes the way of thinking and management technique to improve a company's competitiveness through thorough elimination of waste, the best quality, shortest lead time for arrival of materials, and least production cost. The TPS is a production system to respond to customer needs in a small lot production of many products such as spare parts and not just for mass production.

As a conceptual system the TPS is a manufacturing system that places focus on the highest quality achievable, lowest cost, shortest lead time, through quantity control to reduce costs by eliminating waste,is built on a strong foundation for stability in process and product quality, is fully integrated, is continually evolving to meet up with realities of manufacturing environments, is perpetuated by a strong healthy culture that is managed consciously, continuously, and consistently through standardized work.

The TPS is built on two pillars that deliver operational excellence in manufacturing, the Just In Time pillar and the Rigorous Quality pillar otherwise known in Japanese as Jidoka. Just In Time (JIT) is speedy supply of exactly the right quantity, at exactly the right time, and at exactly the correct location. JIT is at the technical heart of the TPS. JIT is more than just an inventory control system; it is a deep understanding and control of material variations. JIT seeks to produce only what is required and no excess, when it is needed, and in the right quantity. To make JIT work, three principles are required: post process pickup, flow process, and tactful determination of the required quantity. These are perfected through the implementation of Kanban, leveling, standard operation, one-piece flow, and first in first out (FIFO).

Jidoka, the second pillar of the TPS are methods for the combined use of machines and manpower, in other words, automation with human override through passionate talent, utilizing manpower for the unique tasks that require a human touch and machines for quality control using transducers, machine learning, and machine vision to deliver desired product quality. If the machines derail and start producing defective products a human operator must shut them down immediately and alert other operators of the abnormality or problem state by engaging continuous learning and willingness to take risks as best practices through deliberate adoption of technology, rapid implementation of decisions, and continuous improvements towards machine-manpower harmonization.

The TPS ensures that operational excellence in manufacturing is built by defining value, mapping the value stream, pull/flow, and striving for perfection through continuous improvement.

Technically, jidoka uses tactics such as poka-yoke, (methods of fool proofing the process) andons (visual displays such as lights to indicate process status especially process abnormalities), and 100 percent inspection by machines. It is the concept that no bad parts are allowed to progress down the production line. This not only is needed to protect the customer and reduce scrap costs, it is a continuous improvement tool and is a key element in making kanban work as it is a violation of kanban rules to allow transportation of defective parts.

The template driven TPS is what gave rise to lean manufacturing. The focus on "Just in Time" material availability and "Flow-Pull and Leveling" are major components in any mass production system.

The underlying objective of lean production is the elimination of waste. In the Toyota Production System, the seven forms of waste in production are:

Defects	Production of flawed parts
Overproduction	Production of more parts than required
Waiting	Workers waiting
Non-utilised talent	Non assignment of redundant workers
Transportation	Unnecessary movement and handling of materials
Inventory	Excessive inventories
Motion	Unnecessary movement of workers

Excess processing	Unnecessary processing steps
Services	Superfluous amenities

Ford's Moving Assembly Line

Another groomed manufacturing system for mass production is the Ford Moving Assembly Line. Henry Ford in December 1913 had installed the first moving assembly line for the mass production of an entire automobile. The immediate impact of the assembly line was revolutionary. Henry Ford's innovation reduced the time it took to build a car, provided that all parts and materials are on ground, from more than 12 hours to one hour and 33 minutes. The moving assembly line was the most significant piece of Ford's efficiency crusade for mass production of automobiles. Henry Ford revolutionized industrial processes by perfecting the assembly line thereby significantly reducing cost, which enabled him to lower the Model T's price from $850 in 1908 to $300 in 1924, making car ownership a real possibility for a large share of the American population.

It is over 100 years since Henry Ford introduced the modern assembly line and forever changed the course of manufacturing. The combination of optionally interchangeable parts and time efficient processes created a system that eventually sold 15 million model T Fords and made the automobile one of the centerpiece of American culture. Over a century later, the modern assembly line is still the main arm of the global manufacturing industry, that quality manufacturing produces quality products, and is still evolving.

The modern day assembly line is still a framework of Ford's moving assembly line template with attachments of automation and robotics as accompaniment. In the modern assembly line, there are options for preliminary processing of raw materials before branching into the assembly line. Global manufacturing of consumer goods and in fact globalisation itself in based on Ford's Moving Assembly Line. More optimised Ford moving assembly lines exist today with heightened speed and accuracy of production, yielding high quality products. Executives of these companies often introduce innovation to reduce waste and implement automated processes. With the risk of poor customer satisfaction, enterprises have so much to gain by investing in and prioritising quality in their assembly lines.

The highflying mobile executives of todays manufacturing concerns demand real time access to manufacturing data. Therefore, they are ever connected to their assembly lines wherever they are for real time decision making through remote monitoring, gathering, and processing of real-time data at will.

For remote control of manufacturing, factory executives integrate Programmable Logic Controllers (PLC) and Supervisory Control And Data Acquisition (SCADA) with their assembly lines. A PLC is a piece of physical hardware connecting local sensors and devices for data acquisition. SCADA, on the other hand, is software that controls the entirety of the system through the PLC hardware, collecting data from all inputs and monitoring all devices. SCADA is already an essential mechanism for smart factories for a more cost-effective control of manufacturing systems. When integrated with a Remote Terminal Unit (RTU), the SCADA ecosystem can implement alarm point control as in a TPS granting priority for human override to attain predetermined product quality.

Ford's model is a time proven method to get mass production right and sustainably deliver cost effective, market resilient consumer products. Henry Ford's use of interchangeable parts, which was ahead of its time as it is preferred in todays customisation of customer orders, allowed for continuous workflow and more time on task by factory manpower. Manpower specialization resulted in less waste and a higher quality of the end product. Implementation of this manufacturing model dramatically speeds up production of quality consumer goods. The assembly line has forever changed the way people work and live, making people live in urban areas around the world to enjoy public utilities, mass oriented economic programs, and concerted efforts to promote sustainable livelihoods, with ready availability of repetitive factory work and low-skilled jobs for economic empowerment.

Further, the assembly line has highly impacted humanity with pervasive industry, innovation, and infrastructure, massively providing decent work to a vast majority of the population, offering economic growth to societies, developing sustainable cities and communities, promoting responsible consumption and production. In summary the assembly line delivered luxury, convenience and freedom to humanity and as such will stay a while longer as the framework of modern manufacturing with an overriding agenda that benefits the generality of humanity, in other words, Ford's moving assembly line advances social progress.

Flexible Manufacturing System

A flexible manufacturing system (FMS) is a production method built on automation that is designed to easily adapt to changes in the type and quantity of the product being manufactured. Very frequently, short run production is the reality of manufacturing and at such circumstances a programmable environment is what fills the bill. Due to backorders in many industrial environments there is often the need to go below economic orders, even though what are involved are quite essential but the order size cannot just be more at that time and there could be a bottleneck on the supply chain without the order being filled.

FMS thrives on the components of processing stations, material handling and storage systems, as well as computer control systems. FMS works like the assembly line too, however with far greater automation. The material flow in an FMS are with automated material handling systems with integrated counting, weighting, and machine vision that knows exactly the state of the material reaching any workstation based on parameterised information in the central database. The processing stations accept materials branching into the assembly line based on instruction sets from the control computer after the material has been certified to be in a state of readiness for the assembly line, which if the state is not met there are remedial actions to be instructed on what preliminary processing to be carried out and the particular workstation where such is specified to be done. When those preliminary processing is complete the central computer is notified while the processed material is kept on hold. It is when the central computer needs the processed material that it gets released on a journey to the assembly line where it is integrated with other materials for transformation.

FMS smart material handling uses heavy equipment and power tools to move materials into the assembly line where they are transformed, sometimes without human intervention. The material handling equally carries out the series of inspections required in a manufacturing system like the TPS and has the option to shut down production if there are deviations from quality expectations set in the central computer.

FMS smart storage system shines by placing fully processed materials on the conveyor belt to move the products to its final warehouse storage. Every component of an FMS integrates a robot to ensure that setting on quality can be sustained in the materials it handles and remedial strategies are built into the central computer as to what actions to take when the robots observe any deviations from quality expectations.

Six Sigma

Six Sigma is a set of techniques and tools for process improvement introduced by American engineer Bill Smith while working at Motorola in 1986 for reducing errors in manufacturing processes. Six Sigma strategies seek to improve manufacturing quality by identifying and removing the causes of defects and minimizing variability in manufacturing and business processes.

Six Sigma strategies seek to improve manufacturing quality by identifying and removing the causes of defects and minimizing variability in manufacturing and business processes. This is done by using empirical and statistical quality management methods and by hiring people who serve as Six Sigma experts. Each Six Sigma project follows a defined methodology and has specific value targets, such as reducing pollution or increasing customer satisfaction.

The term Six Sigma originates from statistical modeling of manufacturing processes. The maturity of a manufacturing process can be described by a sigma rating indicating its yield or the percentage of defect-free products it creates specifically, to within how many standard deviations of a normal distribution the fraction of defect-free outcomes correspond.

Six Sigma in manufacturing is goal directed. Adopting Six Sigma implies that there be a continuous effort to achieve stable and predictable process results, that manufacturing and business processes have characteristics that can be defined, measured, analyzed, improved, and controlled, and that achieving sustained quality improvement requires commitment from the entire organization, particularly from top-level management.

The six major tools of Six Sigma are: continuous improvement; culture of quality; lean; process management; root cause analysis (RCA); statistics; and value stream mapping (VSM). While the five stages of Six Sigma are: define, measure, analyze, improve and control, popularly shortened as DMAIC.

Companies that implement the Six Sigma model are known to have the lowest defects in their industrial sectors. Companies with a Six Sigma programs are known to be above in terms of quality, cost, and customer satisfaction. Six Sigma is relevant in today's manufacturing business due to its robust principles such as: sharply focusing on the customer; usage of data and analytics to make informed decisions; and the ROI language of management.

Production Related Digital Skills

Manufacturing makes the world go round. The manufacturing ecosystem today is an integrated, communication and computer-based system, with centralised dashboard controls of virtual reality simulations with distances not a barrier. Skills gap required in manufacturing are consistent with digital realities of the day as smart manufacturing take manufacturing into the realms of digital realities. The broad skillset mean manufacturers have a steep learning curve to contend with but those likely to enjoy the journey are those who have already embarked on a digital journey with digital fluency and are at home with digital ecosystems. Most scalable machines for the assembly line these days are programmable, with computer interface, and able to be monitored remotely.

Evolution of manufacturing continues to push the envelope of consumer goods with manufacturing innovation for low cost alternatives to regular consumer goods. As manufacturing becomes mainstream to a lot of interests more innovation will be witnessed but addressing manufacturing the present limited number of platforms of mass production, Henry Ford moving assembly line, the Toyota production system/lean, flexible manufacturing systems, or the Six Sigma, and perhaps a new framework in the making right now.

The expanding envelope of manufacturing is pushing manufacturers to explore new ways; routings, tools, and infrastructures to deliver the big bang of consumer goods to satisfy changing consumer demands.

The ecosystem of manufacture; the demand chain, supply chain, the logistics, and infrastructures are all drastically changing in favour of digital alternatives capable of scalable realisation of the same objectives. These changes are causing disruptions in manufacturing that is unlikely to reach steady state soon.

Manufacturing is going digital with smart factories looking like the future trend of manufacturing, additive manufacturing making a printable black box out of manufacturing shop floors, digital marketing making money out of social media, smart order fulfilment fetching higher numbers through customer satisfaction, supply chain management, virtual requisition, product life cycle management (PLM), quality control (QC), shop floor management, and work force management (WFM) all team up as great digital skills to produce the goods that customers crave to buy.

Personalisation of manufacturing is forcing manufacturers to build to order rather than build to stock placing more demand on the supply chain and forward logistics. In the same vein environmental concerns are prevailing on manufacturers to kick reverse logistics into operation for recycling expired consumer goods rather than let them pollute the fragile environment, extending the life cycle of those products for economic rejuvenation.

Digital Marketing

Digital marketing is by far the most important digital skill for the manufacturing environment; with it there is certainty that products will not pile up in the warehouse. Digital marketing empowers the customer on the internet with customer touch points to enjoy enterprise brands. Digital marketing interfaces the customer to the critical handles of the enterprise revenue cycle and customer satisfaction management systems. These are both flourishing for the enterprise, first it oils the company's activities, and second it sends the company in the right direction of brands to sustain. Enterprise customers are to be acquired with digital marketing, by spreading the enterprise through social media, where all you need is a phone number, then an email, down the sales pipeline you want to know what the customer experiences and how she objects to your brands. Need satisfaction selling kicks in, objections are handled, the customer is enhanced, the first order is taken, a cross sale is made, and the customer is retained, then up sales follow. An annual estimate of sales revenue from the customer looks good, more customers come in through word of mouth referrals, business booms and so do customer satisfaction boom. Digital marketing is that customer tool that powerful forward thinking enterprises use in their CRM arsenal to perpetually stay in touch with their customers

Integration of digital marketing with Enterprise Resource Planning (ERP) unlocks relationships and resources of the enterprise to benefit from top notch marketing that delivers decent CSAT numbers, converting customers, patronising beneficial vendors, recruiting top shelf talents, handling robust finances, sequencing valuable production systems, and sustaining best-in-class warehouse storage systems for value retention. The integration of e-business interfaces three resource ingredients and three relationship communities which are financial resources, production resources, and warehouse resources, as well as customer relationships, talent relationships, and vendor relationships.

Digital marketing provides a cost effective option for the acquisition, enhancement, and retention of customers that keep all ERP relationships and resources working at optimal yields to harness greater opportunities from the demand chain, the supply chain, and for the best options in value creation. With proven strategies delivered through digital marketing the Eldorado of realization of e-business being value creation as a low hanging fruit of globalization. When enterprises achieve brand reputation through digital marketing they are able to better galvanize the resources and relationships of their enterprises through full implementation of ERP solutions where all operations enjoy the benefit of leveraging automation through cloud based processes.

Integrating digital marketing as the cornerstone of e-business strategy is a cohesive online strategy for business unification, operational optimizations, efficient backend collaboration, and flexible revenue, expenditure, and conversion cycles. If all these work together the enterprise gets a handle on basic ingredients of customers, finance, materials, talents, vendors, and storage processes, mechanisms, talents, and infrastructures.

Taking broadcast media and push communications out of the marketing mix results in huge savings and better manageability of technology backed business innovations that offer cloud based solutions that reach the far recesses of the internet to make business goals easily evaluated and achievable, while persistently discovering new ways and employing cutting edge strategies for today's enterprise. The sheer wastage that enterprises are stacking up against themselves by way of campaign aggregates based on old media mindsets are some of those elements plunging major economies into recession.

What digital marketing integration with ERP brings to the table is enterprise immersive experience across all teams and able to deliver steady results based on the traffic it builds from the extensive Internet marketplace.

With the burst of digital marketing taken beyond campaigns a lot of productivity and yield assurance is guaranteed by ERP platforms for enduring productivity beyond the promise of marketing to sustain and deliver the enterprise core business.

Order Fulfillment

Still a CRM plug in too, order fulfillment interfaces customers to enterprise products with delivery strategies. Significantly as an enigmatic tool that takes surveys at the most appropriate in a customer lifecycle-during order delivery. This is the time when the customer is asked what level of satisfaction he feels with the enterprise (5 very satisfied, 4 satisfied, 3 neutral, 2 dissatisfied, 1 very dissatisfied).

Inventory Control

Materials are what to get transformed in manufacturing. Lots of materials are required to produce consumer goods, and they must be held in warehouse before commencing production. Inventory control accounts for movement and value of materials so that they can sure to be available when needed for production and that at the right value as well. Inventory control in most cases is for level monitoring, to monitor stock levels for triggering actions for response by the supply chain, the fulfilment team, the requisition crew, and the replenishment crew. Depending on the replenishment pattern of inventory, there is an economic level of holding for central warehouse, the shop floor, or even storefront. It is the goal of inventory control to maintain data for all holdings and satisfactorily sustain their preferred levels.

Inventory is brought in through the expenditure cycle by issuing purchase orders to vendors from the procurement system. They are taken out through the revenue cycle based on customer orders acted on by the fulfilment system. They are created through the conversion cycle by executing production orders generated by the master production schedule (MPS) for make-to-stock (MTS) production orders.

In production the MPS is accompanied with bill of materials (BOM) of product structure and fed into the material requirements plan (MRP) to determine gross requirements and then applied to the on hand quantities for net requirements. MRP as a source for inventory movement, computes material requirements needed for production and based on this the procurement system places orders with vendors for the MRP net requirements to complete the production order from the MPS.

A popular way to manage inventory is through the reorder point system. In the reorder point system inventory levels are managed such that a maximum level and minimum level are set for materials to be sustained with this strategy. When the inventory kicks of from a maximum it's consumption forms a downward slope in time to the right. Consumption draws down the on hand quantity, when the on hand quantity is equal to the predetermined reorder level a trigger is kicked that sets up an order to the preferred vendor for the material on file for a reorder quantity. The vendors lead time is a factor as the lead time must be accommodated within the safety period for the material on file which is the time in days between the time the reorder level is reached and the time the minimum level is expected to be reached.

The reorder point system being a geometrically derivable system is based on an economic order that period definitions be made at which enough materials are held in stock to address consumption anticipations based on which a product of period consumption quantity and maximum allowable duration gives the maximum quantity of items required to be held in stock. The minimum item quantity is a product of the period consumption quantity and safety period. The reorder level is a product of the period consumption quantity and sum of lead-time of the vendor and safety period. The reorder quantity to be ordered from the vendor is a product of period consumption and difference between maximum allowable duration and safety period. The reorder point system is considered a saw tooth pattern because of its progressive pattern of a combination of sustained inclination due to consumption and vertical rise due to vendor delivery.

Supply Chain Management (SCM)

SCM as a subsidiary manufacturing process is that process that brings in needed materials into the conversion cycle of the manufacturing system for transformation and value addition to finished products. SCM must be practiced as a digital skill with the full spectrum of virtual replenishment, material handling, and logistics.

Enterprises with an instantiated supply chain build rich galleries of vendors by acquiring and retaining the best capacities to derive optimal value from their procurement channels. By integrating capacity planning through material ordering, warehousing, and logistics, manufacturers can take the long haul of daisy chains that their shop floors need to thread the complete consumer goods

Virtual Requisition (VR)

Virtual requisition (VR) empowers vendors to move materials to warehouse in anticipation of value addition to support the conversion cycle and offer the economies of scale of materials to the enterprise. Materials are key to a swift assembly line. However, warehousing materials demand they be controlled by an economic order quantity with a parametric tie back of reorder levels, reorder quantities, and safety periods for every significant material engaged in production. These parametric provisions are vendor driven at most times and often fail. However, empowering the vendor as a stakeholder that gets fed first hand on hand quantities of materials compels vendor commitment to the ideals of advanced shipment notification (ASN) as a way to commit his patrons of state of materials in transit for necessary tracking.

VR grants vendor access to reorder level data on agreed materials to expedite replenishment at established reorder quantities whenever reorder is triggered or specified materials, followed with Advanced Shipment Notice (ASN) that provide detailed electronic manifest of shipment content stating transportation mode, carrier and expected delivery date for stakeholders to receive early notification of shipments ahead of inbound logistics with a tracking handle.

The customer benefits from low ordering costs and lead time reduction while the vendor benefits from orders received on time and possibility to share critical information about materials and their alternatives as well as fulfillment status through ASN and their value chain significance helping the vendor aggregate material requirements for economies of scale.

VR done right personalises the supply chain while according benefits of economies of scale on materials helping to minimise material costs for greater enterprise margins.

Product Life cycle Management (PLM)

A product life cycle is the amount of time a product goes from being introduced into the market until it no longer commands shelf space appeal. There are six stages to the product life cycle: development, introduction, growth, maturity, saturation, and decline. Newer, more successful products push older ones out of the market. The product life cycle describes the period over which an item is developed, brought to market and eventually removed from the market.

When a product begins its life cycle, it may have little-to-no competition in the marketplace. Then, if it does well, competitors may start to emulate its success. The more successful the product becomes, the more competitors it will face. This may cause the product to lose market share, eventually leading to its decline. Production must coordinate its manufacturing runs with the appropriate stage of the prevailing product life cycle so it can give it the right attention through the shop floor with an understanding of what customer expectations to be mindful of for the best marketplace response.

The management of product life cycle through a succession of strategies enables the enterprise deliver the best customer experience. Products have a limited life. Its marketing must address the challenges that are consistent with current stage of its life cycle. Products require different management response at each stage of their life cycle. Efficient management of the product life cycle is therefore required to make it valuable for its intended market.

Enterprises face competitive rivalry from the market in the process of product placement. Therefore product survival depends to a large extent on its timing and strategic response to the rivalry it faces. The winning dominant response of a successful product entry is derived from its flourishing marketplace agility based on its lifespan. However, options exist to extend its lifespan if seen to be dwindling. These options include: rebranding, price discounting and seeking new markets. These options keep the product in the maturity phase of its lifecycle rather than going into decline.

While the revenue and expenditure cycles are important to accountants, the PLM is more useful for the marketing department. It helps the marketing team decide when it is a good time to advertise, reduce prices, explore new markets or create new packaging.

The conversion cycle follows a fairly standard path. First, a product idea is introduced, and then sent to research and development to determine the product's feasibility and potential profitability. Next, the product is produced, marketed and rolled out. This is called the growth phase of the product. If the new product becomes successful, production will increase until the product becomes widely available and matures. This is called the maturity phase of the product. Eventually, demand for the product will decline, and it will most likely become obsolete, resulting in the decline stage. Understanding a product's life cycle is important for a successful company.

The way a company markets a product depends, in part, on its stage in the product lifecycle. A brand-new product, for example, must be explained to consumers. A product that is further along in its life cycle will need to be differentiated from its competitors.

Quality Control (QC)

Quality Control is a major digital skill to be possessed. Since production that makes it to lime light has to be through mass production and there is less and less time allocated to every product coming out of an assembly line. Therefore, the key to good modern production has to be quality control (QC) to get it right. QC is a process that seeks to sustain product quality all through production, into warehouse storage and through product distribution. To achieve the best quality, QC ensures there are testing units to determine that quality is within the bounds of quality specifications for the final product where quality is rigorously tested through product progression and all through to customers perception of the highest quality standards attainable. QC seeks to enhance product quality and reduce risks, gain production efficiencies, and retain customer loyalty.

For consistent QC in a manufacturing establishment a quality control system is setup to manage quality. The purpose of a quality management system (QMS) is to ensure every time a process is performed, the same information, methods, skills and controls are used and applied in a consistent manner. If there are process issues or opportunities, this is then fed into the QMS to ensure continuous improvement.

QC Professionals often attain ISO 9000 certification to enhance their acceptability as experts in their respective QC fields of endeavour. ISO 9000 is a set of international standards on quality management and quality assurance developed to help companies effectively document the quality system elements needed to maintain an efficient quality system. They are not specific to any one industry and can be applied to organizations of any size.

ISO 9000 provides 8 quality management principles to be used as a framework and guideline towards improved performance as:

Customer Focus. Organisations need to retain their customers and therefore should understand current and future customer needs, should meet customer requirements and strive to exceed customer expectations for increased revenue and market share obtainable through flexible and fast response to market opportunities; increased effectiveness in the use of the organisation's resources to enhance customer satisfaction; and improved customer loyalty leading to repeat business and customer retention.

Leadership. Leaders establish unity of purpose and direction of the organisation. They should create and maintain the internal environment in which people can become fully involved in achieving the organisation's objectives for motivation towards the organisation's goals and objectives; evaluation, alignment and implementation of activities in a unified way; and minimisation of miscommunication between levels of an organisation.

People Involvement. People at all levels are the essence of an organisation and their full involvement enables their abilities to be used for the organisation's benefit towards motivated, committed and involved people within the organisation; innovation and creativity in furthering the organisation's objectives; people being accountable for their own performance; and people eager to participate in and contribute to continuous improvement.

Process Approach. A desired result is achieved more efficiently when activities and related resources are managed as a process for lower costs and shorter cycle times through effective use of resources; improved, consistent and predictable results; and focused and prioritised improvement opportunities

System Approach to Management. Identifying, understanding and managing interrelated processes as a system contributes to the organisation's effectiveness and efficiency in achieving its objectives towards integration and alignment of processes that will best achieve the desired results; ability to focus effort on the key processes; and providing confidence to interested parties as to the consistency, effectiveness and efficiency of the organisation.

Continuous Improvement. Continuous improvement of the organisation's overall performance should be a permanent objective of the organisation for performance advantage through improved organisational capabilities; alignment of improvement activities at all levels to an organisation's strategic intent; and flexibility to react quickly to opportunities.

Factual Approach to Decision Making. Effective decisions are based on the analysis of data and information towards informed decisions; an increased ability to demonstrate the effectiveness of past decisions through reference to factual records; and increased ability to review, challenge and change opinions and decisions.

Mutually Beneficial Supplier Relationships. An organisation and its suppliers are interdependent and a mutually beneficial relationship enhances the ability of both to create value towards increased ability to create value for both parties; flexibility and speed of joint responses to changing market or customer needs and expectations; and optimisation of costs and resources.

Shop-Floor Management

Manufacturing requires the efficient and productive running of shop floors to deliver quality products through effective material handling strategies with proper routing, effective timings, machine management, and wastage elimination. Today's shop floors are automated and most machines programmable and with computer and networking connections.

Shop floors are to be managed for safety, quality, delivery, and cost. The machines and tools of shop floors pose a risk to manufacturing manpower and should be prevented from causing expensive lost time injuries. The value addition process of the shop floor should be scrutinized to ensure that the highest standards of quality are sustained in production.

Automated material handling strategies that move materials through conveyor belts must expedite deliveries of WIP from various work centers and rapidly move finished products to warehouse storage at completion of production.

Aggressive cost reduction strategies must be implemented on shop floors to eliminate wattages and sustain cost cutting to deliver least cost production.

To ensure seamless machine, manpower, material and method management of shop floors, there are shop floor management systems that integrate these elements in a GUI dashboard that provides metrics of a manufacturing environment at the fingertips of shop floor supervisors.

Workforce Management (WFM)

Workforce management (WFM) is an integrated set of processes that an enterprise uses to optimize the productivity of its workforce of manufacturing manpower. WFM involves effective manpower forecasting in its labor requirements and creating, dispatching, and implementing manpower schedules to accomplish work schedules and production execution on a day-to-day and hour-to-hour basis.

Activities needed to sustain a productive workforce, such as recruitment, human capital management, organizational behaviour management, and employee compensation must be robustly coordinated. These are geared towards enhancing productivity and performance for growth and adaptation to marketplace requirements. WFM raises the index of competencies with motivation through learning and development. In today's workforce management organizations sustain relationships with talents to sustain a rich gallery of talents in their enterprises by acquiring and retaining the best minds for optimal contribution to enterprise success. The mechanisms to grade and streamline employee capacities for compensation and productivity for enjoyable and profitable engagement for best enterprise culture builds out the enterprise barebones for best of breed structures.

Mechanisms are put in place to secure the health, safety, and working environment for employee comfort to sustain enterprise growth and adaptation.

The performance boosts required in WFM is for time management and waste elimination. By delivering motion studies, being a systematic way to determine the best method of doing work by scrutinizing the motions made by the worker or the machine, of the manufacturing environment optimal methods for performing each task as efficiently and as safely as possible helps to eliminate bottlenecks that saves time, freeing up unnecessary working routings that causes wastages.

Digital Production Strategies

The future of manufacturing is pervasive; it is digital, disruptive, and universal. The world is resource intensive and a small fraction of the worlds resources need to be converted to consumer goods to support the lifestyle of the global population. In making the products the consumer needs there is less and less boundary between manufacturing and the customer because customer demands are more sensitive than before and so do the goods the consumer demands from the manufacturer. This brings the consumer face to face with the manufacturer all through the product lifecycle so the manufacturer can get it right, and the manufacturer cannot afford not to get it right because the complexity of product integrations, the myriads of inputs, the support network, the cross products and in fact a most overstretched value chain that must be sustained to deliver new consumer goods.

The economy of manufacturing demands that new strategies be put in place to support the new manufacturing ecosystem that envisions space age manufacturing while advocating inclusive and sustainable industrialisation and efficient production practices.

New manufacturing shifts are witnessed frequently that make traditional manufacturing models obsolete. The right manufacturing strategy must be technology immersive, time saving, cost cutting, market viable, and value delivering. Increasing product quality and functionality for consumers make manufacturers the generators of value and consumers the stakeholders that profit from values created.

New marketplace realities now accommodate new agile entrants who can easily adopt a confluence of technologies to create value and distribute these values to a technology hungry global community where committed innovators drastically influence the move of smart products from introduction into a prolonged growth and maturity phases by their mere rally of support.

Rapid Prototyping (RP)

Manufacturing is a reality if a presentation of what is to be produced is made. Modelling is required to venture into or commence manufacturing, particularly to convince stakeholders. RP in design thinking and manufacturing is that modeling strategy that involves the speedy creation of a physical part, testing it, and evaluation of a prototype scaled down model of the fabrication. Testing a prototype facilitates information gathering from users, as feedback to improve the prototype, which is advantageous as such feedback would ordinarily have plunged the product into limited marketplace performance if it were a live product. Evaluation determines the quality of the prototype and whether the soundness of the feedback so far acquired warrants the readiness of the iteration of the prototype for product roll out. RP reduces the likelihood of costly mistakes. Production itself could commit to short run multiple prototypes for quick turnaround of the product development process for a more exhaustive feedback mechanism to reach a critical mass in the lifecycle of the product to ensure more potential customers were reached demonstrative of a better product version to be realised.

Further refinements enable developers to accurately capture product expectations and build in the right values before freezing the prototype after all simulations have been addressed. Even after product roll out if new features are desired, because requirements were poorly understood during testing and feedback, the starting point could still be a previous throwaway prototype from where evolutionary constructs are built even though it had previously been frozen.

Evolutionary prototyping builds robust prototypes and constantly improves it, straddling the ground between evolutionary prototyping and staged product roll out. Most prominent parts must be designed first. Evolutionary prototyping addresses risks early, produces steady signs of progress through constant improvement, and useful for rapidly changing customer requirements. However, close customer involvements are required, and could spell trouble for inexperienced modellers, due to scope creep, design decisions, time management, unrealistic performance expectations, and hard to estimate completion schedules.

As experimental and throwaway as RP is it could be expensive to maintain it. It is not all feedback; some feedback could be unconnected, disjointed, and vague. Unworkable feedback must be identified early, and if possible customer could receive guidance capable of enhancing their feedback capacity with morale boosters as part of the prototype itself, since the prototype is to be washed out anyway, at least eventually.

3D Printing And Additive Manufacturing

In 1981, Hideo Kodama, a Japanese, introduced an innovation to RP. He built a layer-by-layer printer using a photosensitive resin that was polymerized by UV light to deliver in 3D solid rather than the traditional flat sheets of paper. The world woke up to 3D printing and since then manufacturers have had 3D printing ink on their fingers, they've all been happily printing their dreams. 3D printing or additive manufacturing makes 3D solid objects from a digital file using additive processes unlike the previous subtractive processes known to manufacturing, where objects are created by laying down successive layers of material, as thinly sliced cross-section of the object, until the object is created. With 3D printing, complex shapes are produced using less material than traditional manufacturing methods. It all starts with a 3D model built in a computer file and transmitted to the 3D printer via Bluetooth, Wi-Fi, flash disk, memory card, USB or by network cable.

Additive manufacturing has transformed the way manufacturing is done. It is now mainstream in most industries. It gets products into the market at prototype speeds. And if it is to benefit prototyping it does it perfectly well and is indeed a major manufacturing strategy both now and into the future as Acumen Research and Consulting forecasts the global 3D printing market to reach $41 billion by 2026.

3D printing solves the problems of speed and lead time, cost reduction, risk mitigation, design flexibility, materials and sustainability.

The 3D printing process is straightforward, the process is involves: 3D model creation, STL file creation, STL file transfer, machine set up, build, part removal, and post processing. The demand for 3D printers is rising, it is ushering many more consumer goods into the market. 3D printers can make new objects and new products very fast. Although there are many different printers available, only nine basic types of 3D printing technology currently exist: Fused Deposition Modeling (FDM), Stereolithography (SLA), Digital Light Processing (DLP), Selective Laser Sintering (SLS), Selective Laser Melting (SLM), Electron Beam Melting (EMB), Laminated Object Manufacturing (LOM), Binder Jetting (BJ), and Material Jetting/Wax Casting.

The three most established types of 3D printers for plastic parts are stereolithography (SLA), selective laser sintering (SLS), and fused deposition modeling (FDM). These technologies have significantly impacted the way businesses, professionals, consumers and educational institutions function due to their adoption of 3D printing. Professionals producing prototypes for others with a quick turnaround time are high demand because they are pushing the envelope of manufacturing to new territories. 3D printing has the ability to save time and costs during the creation of prototypes, make highly accurate parts that assist in product development and production, and produce finished products that may be sold directly to end-users.

Among the items made with 3D printers are shoe designs, furniture, wax castings for making jewellery, tools, tripods, gift and novelty items, and toys. The automotive and aviation industries use 3D printers to make parts. With on-demand production, 3D printing allows manufacturers to satisfy the demand for certain parts, without having to consider the high costs of mass-production, rather manufacturing can focus on small, highly customizable batches. This makes the manufacturer more agile. Furthermore, it enables a co-creating process between customers and manufacturers, resulting in a customised products and a making a prime time out of make-to-order type of production.

The main advantages of 3D printing are realized in its Speed, Flexibility, and Cost benefits for small prototype modeling and small batch production runs. For small production runs, prototyping, small business, and educational use, 3D printing is vastly superior to other industrial methods. In manufacturing, 3D printers generate less waste by using a little less than the amount of material necessary for the product eliminating completely the process of drilling, cutting, and milling. 3D-printed manufacturing reduces overall energy waste and has smaller carbon footprint.

Computer Aided Design (CAD)

Computer-aided design (CAD) is the creation of computer models defined with geometrical parameters. These models could be 2D shapes or 3D shapes and they need to be saved in a computer file format. The CAD file could then be applied to CNC machining. Computer-aided design and drafting (CADD) refers to creating designs and schematics in a software environment used to manufacture products. It is, by no means, a new technology. Digitalized design and drafting for production has increased productivity in design compared to drafting with paper and a pencil. Through CAD, different parts of the same product can be created separately and combined in the final stage. These individual parts are saved and are available for reuse later. It can also automatically generate detailed drawings and bills of materials for the manufacturer.

The beginnings of CAD can be traced to the year 1957, when Dr. Patrick J. Hanratty developed PRONTO, the first commercial numerical-control programming system. In 1960, Ivan Sutherland of MIT's Lincoln Laboratory created SKETCHPAD, which demonstrated the basic principles and feasibility of computer technical drawing.

In CAD the CAD software installed in a computer is fed with design problems. The CAD software possesses smart algorithms for delivering a diverse set of favourable solutions in the problem domain with optional model templates in 2D and 3D to which the human operator makes choices based on objectives and goals to deliver the most optimised solutions. Some CAD systems go a step further to integrate a knowledge base and neural networks that lets them bring up many more optional solutions from which the human operator can choose from a set of best in class solutions that have been field proven by other operators in similar domains, thereby shortening the time required to derive a solution.

The CAD system's knowledge base is also able to guide and advise an operator as to how best to commence particular designs using extensive industry specific rule algorithms that best fits the design at hand as well as design check to ensure the design conforms to predetermined guidelines. In modern CAD designs can be generated from as much as concepts and ideas inputted by the operator, then the CAD software rapidly delivers the design based on its parametric coordinate geometry systems while the operator just sits back and watch. Subsequently, the operator modifies the design until it becomes a 'best fit' or a 'fit for purpose' and rounds it up with a save to file or cloud, print to paper, or send by email for further deliberations or processing.

Modern CAD features interactive modeling, automated analysis, designer views, and drawing creation. A designer view supports multiple design codes and standards. The designer view also includes a wide range of standard drawings and reports to help designers create their own real time customized drawings and reports. CAD has come a long way. It is no longer in the domain of light pens, graphic tablets, and sophisticated VDUs. CAD is now in the domain of ideas. There is a CAD anywhere ideas are thriving, be it smartphones, tablets, laptops, or desktops.

Computer Integrated Manufacturing (CIM)

Further into manufacturing with computers, apart from flexible manufacturing system, there is also a strategy of Computer-integrated manufacturing (CIM) where computers are integrated in the control loop of entire manufacturing processes so that the processes talk to computers rather than being under human supervision. The computers sequence, pause, and halt the processes to ensure materials are adequately integrated into the assembly line and the output of each process meet predetermined quality standards. Manufacturing with CIM is in most cases done with robots or CNC machines, rather than just humans, that receive sequencing instructions from their supervising computers and perform very well non stop according to the installed software running them. CIM has specific actions that are designed to achieve structural change in critical areas of economic activity for best in class manufacturing.

CIM is an implementation of information and communication technologies (ICTs) in manufacturing implying that there are at least two computers exchanging information. CIM operates functional areas of manufacturing, such as design, analysis, planning, purchasing, cost accounting, inventory control, and distribution and are linked through a special computer known as remote terminal unit (RTU) with factory floor functions such as materials handling and management, providing direct control and monitoring of all manufacturing operations.

This integration allows individual processes to exchange information with each part. Manufacturing can be faster and less error-prone by the integration of computers. Typically CIM relies on closed-loop control processes based on real-time input from sensors. It is also known as flexible design and manufacturing.

CIM as a manufacturing strategy eliminates downtime due to insufficient materials as materials will have been preordered from a preferred vendor with an agreeable VR for materials and ASN for order tracking in the CIM database with a measure of reliability, also the need for manpower to devote time for programming machines are non existent as the required machine program is embedded in the central computer and are uploaded at will to the machines so they can most appropriately understand the instruction sets issued from the central computer.

Machine setup time is also at an advantage with CIM as the database of the central computer has on file the setup algorithm required for almost all machines on the planet with options for altering relevant parameters in real time at the time of its production. CIM lets you efficiently manage BOM data, routing sequences and manufacturing specifications by benefiting from built-in best practices. The solution provides a secure platform by centralising all data in one knowledge base to control all machines and computer workstations.

CIM provides cost reduction benefits as accurate information handling saves manufacturing time significantly. The degree of integration that CIM offers enables the flexibility, speed and error reduction required to compete for marketplace leadership. CIM gives greater control of the production process so it can be fully automated to derive the benefits of economies of scale.

Material Requirements Planning (MRP)

A major manufacturing strategy that is essential in materials management is MRP. MRP takes the material teardown of manufacturing to great heights by integrating what needs to be made with how the product is formulated, more of building from first principles, being from the material structure of the products. MRP is a universal manufacturing strategy because it takes a master production schedule (MPS) and a bill of materials (BOM) into its input to project the materials required for producing the independent items stated in the MPS to derive the dependent items being what needs to be issued by the warehouse to produce what is stated in the MPS.

Being at home with MRP is the secret to efficient inventory control in manufacturing environments and for having the best gallery of vendors from the supply chain to strike flourishing bargains on procurement contracts for encumbrance of materials for manufacturing based on the stated material formulations for each product.

MRP helps production planning, scheduling, and inventory control. In production planning MRP helps to derive the material netting for production. Netting is considered a product of the formulated quantity and the production quantity less on hand quantity of each item of production. This figure is a plan of quantity required if the production is to be embarked upon. In scheduling MRP netting becomes an order to the warehouse to encumber those materials in the inventory for production and where they are not available in sufficient quantity a back order is expedited to the preferred vendor for the material by the fastest means possible, preferably by VR as a speedier alternative to deliver the required materials by a stated quantity. Besides, scheduling production by daily or weekly cycles help to manage the MRP according to installed capacity in such a way that there are less breakdowns and less bottlenecks due to machine overloading. In inventory control MRP guards against stock out of critical materials required in production by ensuring their sustained availability to prevent production shutdown.

When planning, scheduling, and inventory are in order manufacturing is able to engage machines, manpower, margins, and methods to deliver the products stated to be delivered at the right quality and quantity. MRP aids the conversion cycle to engage value addition favourably with the right material counts.

Manufacturing Resource Planning (MRP II)

A major strategy after MRP is the elevation to MRP II, while MRP provides a detailed strategy for materials, there were still a lot to be done about manufacturing resources and that is where MRP II steps in to elevate the benefits of MRP in materials to manufacturing. MRP II after picking what are to be manufactured beyond just tracking inventory that MRP delivered reliably.

MRPII strategy is an MRP makeover to walk the ropes of manufacturing beyond materials management to the other essentials of manufacturing resources, such as cost cutting, expediting manufacturing deliveries, quality control, and machine scheduling. MRP II practitioners are many and understand the language of manufacturing as well as the various MRP II software as the word processor for manufacturing processes. MRP II is a strategy every manufacturing practitioner needs to understand because it embodies the necessary content for modern manufacturing competencies. The integration of data in an MRP II system greatly advances manufacturing.

MRP II inputs are from the MPS, inventory files, and BOM. With these manufacturing is scheduled, material requisitions are scheduled, capacities are sustained, costs are cut, production sequences are sustained, customer deliveries are fulfilled, and material purchase orders are released.

Fourth Industrial Revolution (4IR)

The world has come a long way in industrialising. With rocket science of the 2nd world war, the world of mega logistics, mega factories, then electricity, motors, and power tools. Factories have been building every consumer product desired, and even more. Enterprises moved rapidly and negotiated into digital corridors, like watching the drying of wet paint, we watched books in bookshops turn to ebooks, taxis turned to ride sharing, yellow pages turned to marketplaces, and record stores turned to music streaming platforms, all through digital transformation. As digital transformation makes a landfall on manufacturing the digital transformation to be witnessed are expected to be enormous.

But the capacity required to deliver the products the world needs continues to grow, and so factories grew even bigger to address capacity requirements. However, when all available technologies and their latest versions are put together, such as computers, communications, 5G internet, and robotics it is noted remarkably that a paradigm shift has occurred. The emerging pattern as from 2016, in the words of Klaus Schwab, founder and executive chairman of the World Economic Forum (WEF) became known as the Fourth Industrial Revolution (4IR). The 4IR is a fusion of latest advances in Artificial Intelligence (AI), robotics, the Internet of Things (IoT), genetic engineering, quantum computing, big data, cloud computing, and additive manufacturing. The 4IR is a new era and an innovation in manufacturing towards digitization.

Of course, the robotization of manufacturing has its evident advantages: increased productivity, efficiency and quality in processes, greater safety for workers by eliminating jobs in hazardous environments, enhanced decision making with decision support systems, and improved competitiveness by developing customised products for marketplace leadership. In spite of these advantages, it is feared that 4IR may introduce inequalities, increase risks associated with cyber security, disrupt core industries, and raise ethical issues.

Through the implementation of disruptive technologies, 4IR is able to sustain for the first time a totally integrated factory automation that gives control to standard desktop computers, smart phones, and tablets. 4IR employs cloud based big data technology and AI, with these the control systems are very flexible with more degrees of freedom derived through the flexibility of self learning knowledgebase that incorporates the intelligence of several machines from around the world each sharing their learning experience with each other through IoT. Being a broad based revolution around all the major technologies of the internet age, 4IR certainly has a way to provide superior handles what was obtainable in the mass production revolution of the 1914s, it will offer mass production of consumer goods in a totally integrated smart factory, but more of the production will be based on make-to-order production orders that can be customizable by the customer at a considerable distance away.

This manufacturing revolution is ushering in new ways for manufacturing innovation through the ways materials will be blended with AI coordinated routing sequences composed from knowledgebase curated from best in class methods delivered by the best minds and time tested to deliver the best product lines consuming the cheapest amount of materials in its transformation.

Enterprises will not just be the usual brick and mortar autonomous entities but connected digital enterprises nodes that swim their lanes in the supply chain due to the value they deliver. Digital enterprises will be recognised for value delivery. Digital enterprises will be valued for their economic graph and how socially and economically viable they become. The connectedness of digital enterprises will be a factor that positions them on the value chain. With their workforce of remote talents they may not exist in any office but literally just in the cloud. This will change the structure of taxation, law and order, and even banking as most will do their banking with crypto currencies.

4IR is a great potential to smartly increase manufacturing productivity. From the first principles of manufacturing such as product formulation and method composition with artificial intelligence based on enormous data drawn from the global big data repository these are going to be intelligently derived to fulfill smart demands built around global dashboards from CSAT numbers based on consumer spending, commodity prices, retail prices, household spending and capacity utilisation rate metrics. Making it easy to produce consumer goods on demand and on the fly.

4IR will lead to value creation. Just altering product formulation variables on the manufacturing simulation panel that controls the assembly line will create new products. The mode of packaging and delivery of products will also add if not create new values. As supply supply chains become connected so also will demand chains be connected and the gap between manufacturers and consumers will narrow, reducing production costs and time to market. 4IR will result in a boom economy for all on the graph of the new corridor.

On a flip side there is intense fear that many may be left behind, many of whom are not used to quickly stepping into new corridors. For such, it might be difficult to carry them along if necessary steps are not taken early to win them over as 4IR innovators in their own ways and incorporate their thoughts into the intelligence of the future. A viable place to start is to provide handles of 4IR thinking for developers of ERP software since most manufacturers speak a common language through ERP and most ERP ecosystems are global in nature. 4IR readiness of ERP software will command a prevailing diversion for the world in terms of what next to be done for its delivery.

Shockingly too, the rush to acquire stocks of 4IR patents makes many question the openness of the 4IR and if it could become prohibitive for enterprises, the revolution requires global collaboration and as such should come with a fiat of openness to encourage volunteering of ideas rather than a rush to secure pecuniary interests. This even more so as developed countries have all national 4IR policies to the detriment of developing countries, in a way locking the developing countries out in the dark.

A way to access the non openness of 4IR is in how many fourth industrial revolution roundtables are regularly held in developed countries compared to non existent 4IR roundtables in developing countries, where in reality the reverse ought to be the case, as global resource custodians, if this cannot be the case then 4IR roundtables should be held in every country of the world in the least.

As the race for the 2030 SDG date with goals and indicators heats up, we have seen no indicators nor metrics to signpost a drift to 4IR implying that a more preferred industrial revolution more popular than 4IR may emerge to submerge 4IR for its lack of popularity among the resource custodians of the world. A new platform with measurable goals, targets, and indicators will hold greater promises as 4IR appears dull and capable of being cornered by a few governments, manufacturers, or institutions.

The challenge for developing countries in the global manufacturing equation will remain how to recognise their abundant talents and the competency corridors to place them. The lack of traceable capacity building tracts towards 4IR does not signify a credible interest in talent resources management apart from the interest to mobilise robots, these appears a weaponization plan suggestive of imperialist intent against human talents that are required to upload their intelligence to robots.

Developing countries face prevailing challenges of hunger, unemployment, disease, electricity, education, sanitation, transportation, and housing. 4IR ought to first address these challenges before embarking on industrialisation for universality and inclusiveness of its expected impact. It is in global interest to pursue the inclusiveness of 4IR as manufacturing touches humanity.

4IR development as is presently constituted is a build up to economic inequality by not carrying the governments of developing countries along. There is risk that most of these countries may resort to formations of economic blocks in an attempt to pursue price-based interests as that may be only option to fall back on the long run.

When this happens the economies of the world would have become more fragmented and this may further compound global economic woes. Whereas 4IR done right as a bottom up global development strategy may actually resolve some of the nagging global economic issues to gain traction and developing countries buy in and sponsorship.

The benefits of 4IR are too overwhelming to be left in the hands of a few members of the global community. Beyond the celebration of a few talking heads the world must find the right handles to move manufacturing into the realm of a globally recognizable pattern to sustain social progress. 4IR means a quantum leap for technology. It must represent an acceptable, sustainable platform for global manufacturing, to keep enterprises on the cutting edge of digital advancement, to deliver opportunities for capacity development, and open access to new market opportunities to fuel growth of the global economy.

The scale of automation introduced by 4IR is totally integrated. Every aspect of industrial automation is addressed; the factory could then be run from a desktop or even a smart phone with all actuators and logic controllers intelligently programmed to respond to multiple devices, the IoT scenario where every instruction counts. Beyond the supervisory computers are further digitisation where ERP interfaces the business to ensure all aspects of the business such as CRM, finance, production, manpower, vendors, and warehousing exercise control on the smart factory as well, resulting in a connected factory, a digital enterprise.

The Digital transformation that enterprises now enjoy on an adjusted technology threshold is pushing the envelope of social progress favourably towards the rise of digital platform-based business models, even beyond the realities of SDG targets. If these progresses are inclusive there is great promise for humanity. But there must be a deliberate framework for continuous integration of all aspects of technological frameworks. Headcount capacity of SMEs will continue to be a concern, particularly in developing countries, due to challenges of unemployment. Therefore, the promises of 4IR must free some of its resources for agriculture, employment, health, education, energy, sanitation, transportation, and housing to appeal to developing countries and their annual balance sheets.

The huge value creation potential of 4IR should be made to go round to promote decent work and economic growth. The capacity of the fourth industrial revolution when weighted on the scale of data sourced by MIT Technology Review Insights from McKinsey & Company that about $4T stands to be created through 4IR by 2025, then it is important to take note where the created value should be channelled in the scheme of the world's social progress.

The race to get into the proliferation of machines must be approached with caution. There must be caution that the world doesn't end up building a Frankenstein that puts the human race at risk. Rather, the goal should be the unconditional unlocking of the hidden varieties of human talents to take talents people onboard as key players.

The world of business will witness more small and medium enterprises resulting to the rise of special enterprises as special purpose vehicles such as venture, autonomous, partner, and linked enterprises to achieve specific investment decisions of stakeholders.

The virtual manufacturing of smart factories may be at a reduced time to market, with enhanced flexibility, and at an increased efficiency, but it has to add value to be meaningful. The world is very much accustomed to products so the products that will really count are those that meet demand. Smart manufacturing with its big data will have to track and trace data around the world to determine credible heartbeat demand before landing such data on smart factories for a top down product driven manufacturing through the components of product life cycle management (PLM), manufacturing execution system (MES), and industrial automation system (IAS) for full-scale reconfigurable smart factory implementation through a portal integration of production planning and production engineering.

The smart factory system tie-down to big data must respond swiftly to ability to meet customer demands, reactions to changing market trends, high number of product variants, and production improvements. These incorporate machine individualisation, quality improvements, and self-learning machines. In this environment, productivity is continuously increased through a consistent connection between information technology domains and machine domains for quality management, energy management, production line monitoring, and process improvements. This saves time, reduces costs, and boosts flexibility. Done right, there must be access to production data, alerts for maintenance, remote operation, and direct device diagnostics.

Global Economic Growth Prospects

Full-scale industrialisation is required for economic growth. Manufacturing is a strategic key to economic development. It's a fact that manufacturing ushers in an era of global growth and innovation. Right now manufacturing is assuming prime mention globally, even among societies where manufacturing is lacking as a policy. As manufacturing assumes mainstream in global economy, more players will emerge and more sectors of economies will get pulled up to grow as well, introducing greater depth of understanding into manufacturing algorithms, such that manufacturing becomes a language of value creation, value addition, practices, methods, models, and objectives. The information centric model of manufacturing has always pulled players to adopt flourishing industrial roles that leave the ecosystem always better. With the long way manufacturing has come so far it has become a global repository that delivers all products humanity needs to sustain livelihoods.

While manufacturing is not an island of operation but is piggy backed on the demand chain from where market strategies are drawn as well as on the supply chain from where materials are drawn and further replenished. With manufacturing helping the world go round warehouse resources abound to support the supply chain with value retention. Industrial craftsmanship perpetually delivers the machines and power tools to perform the heavy lifting of industrial workloads. Workmanship abound of the all time high population of the world as the blue collars, white collars and high fliers to deliver the required manpower and talent to mill the assembly lines and creatively drive global economies into realms of flourishing productivity.

Industrialisation is never given to chance. There must be deliberate effort to put infrastructure in place, reform policies and set up roundtables to chat the way forward for contracts to be signed and stakeholders to take up roles so that there could be a step out from traditional economy into the virtual world of digital manufacturing as it is today. Industrialisation involves more than just a cursory mention because when it acquires escape velocity it takes full-scale national engagements to get the machines milling. Where manufacturing does not get off the ground there is the risk of premature deindustrialisation as the muscles, boots, and blue collar do not adequately deliver the economic promises of industrialisation.

Product formulation will become a major proprietary of the industrial age as stakeholders that are creatively adept at concept delivery will keep machineries busy on the platform of ideas which they have the option to instantiate at a fee.

Economic Productivity

A flight of manufacturing is a step into value creation and addition but has to be achieved through interaction with critical information and communication due to the structural factors of competitive rivalry, inherent bargaining power of the supply chain, installed bargaining power of the demand chain, and threat of substitutes. These competitive forces compel manufacturers to repackage their products to achieve a competitive appeal that enables it travel the world. The competitive strength of manufacturers in addressing these forces can vary from one industry to another enabling industries attract compelling returns from their investments based on how competitive forces are addressed.

The productivity achieved through manufacturing is for its contribution to the economy that there is further support sustained across wider sectorial linkages through indirect and induced multiplier effects. Based on the usually favourable ratio of outputs to inputs manufacturers work hard to reduce costs so as to increase productivity through introduction of technology and innovation. While striving to achieve greater manufacturing throughput quality must be maintained to sustain productivity for deeper penetration of related value chains to sustain the harnessing of more diversified opportunities in secondary and tertiary sectors of the economy.

Strategies abound to maintain manufacturing productivity, such as employee up skilling; maintenance culture; workflow review; waste elimination; monitor capacity utilisation.

Up skilling employees through training and development delivers a skilled and knowledgeable workforce essential to productivity, they will deliver to a higher standard with fewer mistakes. Having a maintenance culture in manufacturing maximises the lifespan of the enterprise asset portfolio, this reduces machine downtime, improves product quality, and increases operational efficiency.

Prospects For Digital Production Resources

The whole essence of taking manufacturing into the realms of digital innovation, DX is to realise the begging global insatiable appetite for value creation and value addition required to take global economies into self-reliance and sustainability.

Workplace communication technology advancements have resulted in greater man machine interfaces requiring greater labour specialisation at the expense of manufacturers. More and more talents are up skilled to meet the requirements of digital production resources that sometimes spin out of industrial orbit into solo entrepreneurship giving rise to the buoyancy of new economies hitherto unrelated to their original economies and in the process creating value, adding value, creating new employments, and even in some cases creating new industrial sectors. That is why nobody goes wrong pinpointing digital production as a poverty eradication strategy.

Dashboard data analytics drive the pathological requirements of best of bread manufacturing by keeping eyes on the thread the path of economic significance follows, employing global value chains into national significance for the promotion of favourable business environment and entrepreneurial culture, ultimately towards delivery of high-quality employment through manufacturing. The driving objective should be to key into and operationalize a global strategy for youth employment into smart work by learning how to make better products at lower costs through digital centric production.

Digitalising manufacturing launches enterprises into the jungle of globalisation. Globalisation is going to threaten any industry, but if industrial blueprints are well implemented from first principles it is possible to work alongside globalisation instead of working against it. If economies and enterprises recognise the value chains they are connected to deep enough and plug into them, they will find globalisation can be their friend and work for them, like offering the prospects for wider markets. Globalisation offers immediate solutions to manufacturers through communication, innovation, and collaboration. As global trade and commerce expand beyond domestic boundaries, product standards are increasingly becoming important. Consumer concerns surrounding quality and safety will determine the product marketplace dynamics.

Smart production plugs into the interconnection between various technologies to deliver higher quality products, improve productivity, cost-effectiveness, time-savings, easy configurations, better understanding, quick response to market demand, flexibility and remote monitoring to claim prevailing opportunities provided by resource-based industrialization to get manufacturing running and the economy plugged into beneficial value chains letting manufacturing benefits transform the various sectors as a catalyst for economic growth.

Smart manufacturing can deliver manufacturing containerization through predictability and dependability with increased speed from concept to realisation to deliver smart digital manufacturing where and when by pushing all plant components and support network to where it is needed for instant economic transformation with all the gains of economies of scale, automation, and fast marketplace plugins. The economic benefits of smart manufacturing containerization immediately plugs into global value chains from implementation, adjusting and stabilising at the pace of the concept rather than the pace of the economy.

Smart industrial policies of industrialisation, deindustrialisation, and foreign direct investments (FDI), contributes to sustainable development, can have many social and public benefits for clean air, lower carbon emissions to achieve economic growth through optimum utilization of resources, modernization, balanced industrial development, and balanced regional development. The concessions provided for in a smart industrial policy promotes agriculture, trade, transport, foreign trade, services and social sectors of the economy to increase employment opportunities, national income, per capita income and living standard of the populace.

The industrial policy of a country is influenced by the ideology and principles of the concerned government. The industrial policy helps the country to be self-sufficient and prosperous by preparing a structure and basis of industrial development. Hence, the industrial policy of government must be well defined, clear and progressive. Moreover, it should be adhered to and implemented earnestly to deliver its promises. A smart industrial policy must vigorously pursue modernisation and balanced industrial development.

As a first there must be sustained effort to diversify and grow manufacturing exports in tandem with the high value potentials in information and communication technologies and then restructure for global competiveness across board, export oriented industrialisation (EOI) employs export led growth to harness more natural resources for which there is comparative advantage and with export substitution move these quickly into global value chains for rapid economic gains so it can move into other sectors of industrialisation, the goal of export substitution is to open domestic markets to foreign competition in exchange to foreign markets while working a favourable balance of trade.

Export-led growth is important for developing economies. First is that export-led growth improves the country's foreign-currency finances. Secondly, increasing export led growth triggers productivity increases, sending exports in an upward spiral cycle.

On the flipside import dependency results in slowdowns because employments are outsourced with imports. This is a major challenge for developing countries that for ease of imports have abandoned manufacturing in favour of imports. The risk is that the value chain required to deliver the imported products are by fiat outsourced to the shore where the manufacturing is domiciled. Full range industrialisation will help developing countries achieve full and productive employment and decent work for all women and men, including for young people and persons with disabilities, and equal pay for work of equal value. The policy objective should be to develop indigenous capacity and manufacturing capabilities towards conversion to an export oriented industry for elevated productivity, quality infrastructures, and higher quality of life for citizens, driving the markets from comparative advantage to national prosperity.

Prospects For Decent Work & Economic Growth

Manufacturing industries because they produce the products they merchandise rather than first having to engage the expenditure cycle and the supply chain to acquire marginal values have higher employment, revenue and output multipliers relative to merchandising and service activities because they are creators of values and as such determine a range of economic variables such as employment, consumer prices, national budgets, inflation, money supply, and exports. Manufacturing also promotes stronger inter-industry and inter-sectoral linkages, sustained productivity, technological development and innovation. Engaging manufacturing is to sustain employment, decent work, and economic growth.

In a rudimentary economy, more people depend on a labour intensive primary sector, but as manufacturing being the backbone of industrial development is introduced the dependence on the primary sector is reduced as jobs are generated that moves people away from primary sectors into job opportunities of the secondary and tertiary sectors, every job created in manufacturing has a multiplier effect in creating additional jobs in the tertiary services sector, and then when exports are achieved because manufacturing produces a lot of goods due to excess capacity, expanding trade and commerce beyond national boundaries brings in much needed foreign exchange that will help to modernise manufacturing, engage innovation, and acquire more machines to expand the scale of production, ultimately eradicating poverty, increasing employment and increasing GDP.

The future of manufacturing lies in knowledge intensive multinational conglomerates that grow though mergers and acquisitions to confront regional and international competition to take the reigns of market leadership. Trade and commerce with conglomerates access established multilateral linkages and regional structures such as NAFTA (the North American Free Trade Agreement), EEC (the European Economic Community), and APEC (the Asia-Pacific Economic Cooperation) to massively move products into markets using the best of competitive manufacturing strategies to deliver to customer requirements, while it is necessary to comply with safety considerations, environmental rules and welfare legislation in product delivery.

The strategic decisions of global multilaterals will be between production or assembly and distribution, noting specifically that value addition is achieved by production, whereas assembly and distribution are mere cost variables, to distinguish what tight rope of economic accountability legacy they are walking on.

Constituting manufacturing into an organized and highly productive, knowledge driven industry will spin off into an engine of employment for decent work and economic development. Most economies the world over are still founded on the secondary sector at most and requisite investment in technology and automation is vital to move their economy into the tertiary sector so as to improve their employment situation, deliver a knowledge driven economy, and address fundamental economic challenges and step into the corridors of the major players of the global economy while at the same time addressing social and environmental sustainability which has become a key issue of global interest.

National economies need to sustain inclusive growth to make their economy work for them. There are targets that must be achieved for this to happen. First, higher levels of economic productivity through diversification into sectors of manufacturing by products must be achieved. Second, expansion must be made into the frontiers of knowledge based creative endeavors and innovation to deliver productive activities around manufacturing. Third, resource efficiency in production and consumption to decouple economic growth from environmental interests for sustainable consumption must be progressively improved. Fourth, full and productive employment and decent work for all persons must be achieved. If all these things are done then a globally competitive economy will have been delivered, supporting local culture and promoting tourism arrivals by creating legacy jobs around tourism. In the same vein, Expanded access to banking and creation of access to multiple financial instruments will support a new financial order necessary to sustain the quality of life achieved within a manufacturing economy.

Most importantly, it is necessary to invest in the people through healthcare, education, electricity, sanitation, transportation, and housing to protect the manufacturing ecosystem to prolong the economic gains of value creation through manufacturing.

www.ingramcontent.com/pod-product-compliance
Lightning Source LLC
Chambersburg PA
CBHW032210220526
45472CB00018B/662